BUILDING THE
BIGGEST

The ships that influenced the conception of this book: *Oasis of the Seas* (pictured) and *Allure of the Seas* are the first passenger ships to exceed 200,000 gross tons.

BUILDING THE BIGGEST

BIGGEST

FROM IRONSHIPS TO CRUISE LINERS

GEOFF LUNN

Dedicated to the life of Olive Lunn,
a kind and caring mother whose sad passing during
the compilation of this book gave me an extra
determination to complete the project in her memory.

First published 2009

The History Press
The Mill, Brimscombe Port
Stroud, Gloucestershire, GL5 2QG
www.thehistorypress.co.uk

British Library Cataloguing in Publication Data.
A catalogue record for this book is available from the British
Library.

ISBN 978 0 7524 5079 7

Typesetting and origination by The History Press
Printed in Great Britain

CONTENTS

ACKNOWLEDGEMENTS

Much of the photographic and informative material used for the compilation of this book has been taken from my personal library, which I have built up over the past half-century. However, I extend my sincere thanks to the following shipping concerns for enabling me to visit their passenger vessels in port, arranging cruises for me and providing invaluable facts, figures and images: Carnival Cruise Line, Cunard Line, Diamond Cruise Ltd, Disney Cruise Line (UK Public Relations, The Walt Disney Co. Ltd), Holland America Line, Norwegian Cruise Line, P&O Cruises (UK) Ltd, Princess Cruises and Royal Caribbean International (including their photographer, sbw-photo. com) particularly for their images of *Freedom of the Seas* and *Oasis of the Seas*, under construction, and the onboard view of *Summit*.

I am also indebted to similar assistance received from the following UK public relations businesses: Biss Lancaster Ltd, McCluskey & Associates, Siren PR and Weber Shandwick Worldwide, and to Clydeport and Porthole Productions for photographs of the *QE2* and *France* respectively.

Last, but certainly not least, I wish to thank several shipbuilding organisations for allowing the reproduction of their illustrations and information: Chantiers de l'Atlantique, Fincantieri, Kvaerner Masa-Yards (later Aker Yards and STX Europe) and Meyer Werft.

As my research is derived from material collated over an extensive period of time, there are a number of firms and individual persons who, through mergers, changes of business interests and other reasons, can no longer be found on record. Nevertheless, I wish to offer my gratitude to all concerned, whatever their present circumstances may be.

In addition to the aforementioned, books that constitute my bibliography are as follows:

John Pudney, *Brunel and his World*, Thames & Hudson Ltd, 1974
Sally Dugan, *Men of Iron, Brunel, Stephenson*, Pan Books, 2005

ACKNOWLEDGEMENTS

L.T.C. Rolt, *Isambard Kingdom Brunel*, Longman Group Ltd, 1957

Ken Smith, *Turbinia – The Story of Charles Parsons and his Ocean Greyhound*, City of Newcastle-upon-Tyne, 1996

Tom McCluskie, *Anatomy of the Titanic*, PRC Publishing Ltd, 1998

John Maxtone-Graham, *The North Atlantic Run, 'The Only Way to Cross'*, Cassell & Co. Ltd, 1972

David F. Hutchings, *RMS Queen Mary, 50 Years of Splendour*, Kingfisher Railway Productions, 1986

Douglas Ward, *Complete Guide to Ocean Cruising & Cruise Ships*, Berlitz Publishing, 2005

INTRODUCTION

It was April 1998 and I had been invited to attend that most prestigious of maritime events, the christening of a new cruise liner. I joined the elite group of guests assembled on the Southampton quayside, taking my seat in front of the QEII Terminal. As I gazed up at the great white wall of a ship that towered over us and, from stem to stern, seemed to go on forever, my public relations colleague, a pleasant young lady, observed: 'Huge, isn't she. But they'll soon be building passenger ships twice her size. It's so hard to imagine.'

It certainly was hard to imagine at the time, but she would be proved absolutely right. Passenger vessels *have* increased to twice that size... and more.

The 78,491-gross-ton *Vision of the Seas* was joining a fast-growing modern fleet operated by Miami-based Royal Caribbean International. Her call at Southampton ensured her recognition as one of the biggest passenger ships to have visited British shores, drawing comparisons with the legendary liners *Queen Mary* and *Queen Elizabeth*, both of which graced the North Atlantic until the 1960s. A hush came over the proceedings as the president of Royal Caribbean, Richard Fain, stepped forward to welcome guests and to enthuse for a moment about his company's gleaming new ship. He then invited his wife to perform the christening of *Vision of the Seas*. A bottle of champagne swung with unerring accuracy towards the ship's starboard side, neatly avoiding a window as it crashed into the pure white paintwork. 'Welcome, good ship, to the modern world of cruising.'

'How times have changed!' I thought to myself. If the innovative engineers of the Victorian age guided the shipping industry from sail to steam, wood to iron and later to steel, then the twentieth-century invention of the computer had taken ship construction to entirely new levels. Massive passenger vessels equipped with remarkable facilities and efficient machinery, capable of meeting the highest standards of safety, can now be built from keel to funnel in no more than two years. Construction techniques have changed beyond recognition, as have methods of ship design and, indeed, the very roles that these floating resorts are asked to play. More will be revealed in the coming pages.

Over the course of the 100 years or so, ocean passenger ships were created for the prime purpose of transporting people between the continents of the world. The affluent aristocracy of the day would wallow in first class luxury whilst emigrants, of whom there were considerably more in number, would travel in third class austerity. The ships operated 'line' voyages, keeping to regular schedules, often carrying mail and cargoes. But in 1958 a Boeing 707 made its maiden flight across the Atlantic, beginning a regular commercial jet airliner service, reducing the crossing from around five days to five hours and knocking the first nail into the coffin of the transatlantic liner. By 1965 only 5 per cent of Atlantic business remained with the shipping companies. Furthermore, the '70s heralded the arrival of the giant Boeing 747 on long-haul routes, taking away valued custom from South African and Australasian-bound liners. No emigrant ship sailed between Europe and Australia after 1977. The ocean-going passenger ship was surely doomed.

The maritime world, however, had not accounted for the expertise of an unlikely ship-owning partnership whose foresight induced the birth of a new era of cruising that would grow into the phenomenon it is today; the fastest-growing sector of the travel industry. Passenger liners that had not already made their way to the breaker's yard were

'I name this ship…' – A bottle of champagne smashes against the side of *Vision of the Seas* as she is christened at Southampton.

'If I ruled the world...' The author pictured at the high-tech cockpit-style controls on the bridge of resident ship *The World*.

converted to meet this unprecedented call for holidays at sea. Soon brand new ships were rolled out in increasing numbers as the demands and expectations of the 'vacationers' grew. Passenger ships were now being planned, designed and built for different purposes. Even a 'resident ship,' created specifically to carry people of not inconsiderable wealth continuously around the world, appeared on our seas. But the size and diversity of these ships have evolved not only through massive advancements in technical know-how over the years, but also through the implementation of stringent safety features whose enforcement emanates from a number of serious maritime disasters, some of which resulted in considerable loss of life.

The most famous of these is undoubtedly the sinking of the *Titanic* in 1912, as a result of which the initial version of a most important new international treaty, founded for the enforcement of safety standards of merchant vessels, was passed. The Safety of Life at Sea (SOLAS) regulations have influenced the construction, design and lifespan of passenger ships ever since, first being reviewed and adapted in 1929, and again in 1948. In 1960 the newly formed International Maritime Organisation (IMO) ensured that SOLAS requirements were substantially advanced, in keeping with new technology within the shipping industry, and these were renewed again in 1974 with a view to simplifying the process of making amendments to the treaty.

But on 7 April 1990 a passenger/car ferry sailing in the North Sea was engulfed by flames, the disastrous consequences of which undid all the IMO's good intentions. The 10,513-gross-ton *Scandinavian Star* had been built in France in 1971 and was travelling from Oslo en route for Frederikshavn in Denmark when she caught fire. According to subsequent investigations, corridors on the ferry were narrow and difficult to exit, there was a lack of sprinkler systems or any other automated fire-fighting system, and materials used in the vessel's interior finishings were combustible and emitted deadly fumes when alight. Additionally, many of the crew were not only new to the ship but were unable to communicate in the many languages of the various passengers. Amid the thick black smoke that was filling the ship's public areas, complete panic ensued, leading to the tragic loss of 158 lives.

Scandinavian Star was, in fact, in compliance with SOLAS regulations at the time of the fire, as a general clause that was still in existence even after the 1974 ruling allowed merchant ships to stay in service providing they adhered to the SOLAS regulations relating to the year of their build. Consequently the ferry needed only to comply with SOLAS regulations from 1960. The IMO wasted no time in pushing through new fire safety standards that eliminated the offending clause and, although these were labelled as 1992 Amendments, owners were given until 1 October 1997 to bring their older ships into line. But despite this additional breathing space, a number of well-loved passenger ships were withdrawn from service, their owners deeming the necessary refurbishments uneconomical.

As I write, a further amendment to SOLAS regulations is looming, due for enforcement in 2010. This has already caused the retirement of one very popular liner, as will be discussed later in this book, and several others are expected to pass over the

After the fire. The *Scandinavian Star*, renamed *Candi*, lies at Southampton awaiting new owners.

dark horizon before long. Once again, the concern of fire safety will be at the core of these amendments, and, as cruise liners continue to grow in size, their passengers can voyage with the comfort and confidence of knowing that every conceivable precaution has been implemented to secure their safety. The obligation of ship-owners to comply with SOLAS, together with technical advancements within the shipbuilding industry and the proven theory that it is more economical to 'build big' has led to the emergence of highly modern and innovatively designed passenger fleets over the past decade or so.

In accordance with my colleague's wise predictions of that April day, a pair of cruise ships measuring in gross tonnage almost twice that of *Vision of the Seas* are now in Royal Caribbean service, whilst not far behind in terms of magnitude, the magnificent *Queen Mary 2* spends a season each year as a transatlantic liner. Yet even these giants are being overtaken by another pair of sister ships built for Royal Caribbean that would surely have been completely unimaginable in 1998. My inspiration for writing this book, they are the first passenger ships in history to exceed 200,000 gross tons, and are veritable floating cities, promoted as offering a third more space than any other cruise vessel afloat and, would you believe, measuring SEVENTY times the size of the first Victorian passenger-carrying iron ship that could call itself the biggest in the world.

BUILDING THE BIGGEST

For the compilation of this book I was able to draw on more than fifty years of experience as a keen student of shipping whilst, as you will soon appreciate, several cruises and almost fifty visits to passenger ships in port have allowed me a more personal view of the changing face of the industry. Sadly, since the terrible events of 9/11 in 2001, shipping concerns have become noticeably nervous about showing off their latest fleet additions to anyone without direct connections to their company or the travel industry, whilst port security has been tightened to unprecedented levels. Nevertheless, I have been offered wonderful opportunities to inspect some of the most exciting new ships of recent decades.

Throughout this book I have attributed a degree of flexibility to the term 'the biggest' by covering not only passenger ships that were or, in the case of current vessels, have been the world's biggest at some point in their careers, but also those that can stake a claim to a version of the title, such as the biggest to sail on a particular route or the biggest to fly a certain nation's flag. Furthermore, I have deliberately attempted to restrict the use of technical jargon and facts and figures to a minimum, although at times their inclusion is, of course, unavoidable. By doing so I trust the book will be as appealing to the land-lubber as it will be to the shipping enthusiast. Some readers may even be amid a cruise, lounging with their feet up, enjoying the warm sunshine, the dark skies of winter, under which most of my words have been written, far from their minds. I am acutely aware that certain passenger ship aficionados, with a copy of this book in hand, may have comments to make concerning the final two words of its title – 'cruise liners'. They will be observing that a passenger vessel that is employed solely in cruising should be referred to as a 'cruise ship', whilst a ship engaged in the ferrying of people across the oceans on regular schedules and, more often than not, built with added strengthening and a deeper draught, is a 'passenger liner'. I will certainly not argue with that, but my extensive coverage of the *QE2* and *QM2* (*Queen Mary 2*), both of which have combined transatlantic 'line' voyages with worldwide cruising, and the fact that many modern cruise vessels are operated by cruise 'lines', must add some credibility to the choice of those two words which, after all, do fall more comfortably within the book's title. I shall say no more.

So join me as I journey through almost two centuries of incredible change, following the development of the ocean-going passenger ship from the days of paddle wheels, auxiliary sails and coal-fired engines, through the traumas of two world wars, to the enormous cruising cities of the twenty-first century.

CHAPTER ONE

A MAN OF IRON

Stories of nineteenth-century Britain, under the influential rule of Queen Victoria, invariably paint a sober picture of ladies' ankle-length fashion and morals as rigid as the tight waistcoats and white collars worn by the men. Yet this was a period of fast revolutionary changes; an era that saw Britain, and eventually other parts of the world, move forward from the days of uneven, potholed highways to coal-fired cross-country railways, from wooden sailing ships to iron-built passenger-carrying steamships.

The Industrial Revolution, which had enveloped Europe since 1760, witnessed the emergence of brilliant innovators who, with their imaginative minds and through relentless hard work, guided Britain into a new age of discovery that paved the way toward today's highly technical scientific world. Among those pioneers was an Englishman with a French name, a man of small physique but, in terms of the sheer diversity of his innovations, head and shoulders above the rest.

Isambard Kingdom Brunel was born in Portsea, Hampshire, on 9 April 1806. The third son of Marc Isambard Brunel, an eminent engineer whose birthplace was a tiny Normandy village in France, and Sophie Kingdom, his English mother, he was quick to follow in his father's footsteps, as, at the young age of four, he was capable of drawing a perfect circle. Following early education at Hove, Sussex, young Brunel was sent across the English Channel to study at the College of Caen in his father's native Normandy, and at the Lycée Henri-Quatre in Paris. But at the age of sixteen Brunel was back in England working at his father's London office, and two years on, in 1824, he became involved in Marc's greatest engineering achievement: the boring of the Thames Tunnel, the world's first underwater thoroughfare to be constructed beneath a navigable river. In little time he was appointed chief assistant engineer of the project that was to become beset by so many problems that it took more than eighteen years to complete. There was a permanent risk of accidents, particularly at high water, and on one such occasion young

Brunel became trapped by a huge surge of stenching, sewage-filled water. It was a miracle that he survived, although over the next seven years he was plagued by ill-health before making a move to Clifton, near Bristol. The rest, as they say, is history.

Isambard Brunel was soon embarking on a series of solo engineering feats, although he was at this time still some way from his illustrious involvement with ships and the sea. He designed the Clifton Suspension Bridge, regarded even today as one of his finest pieces of work, a masterpiece spanning more than 213m over the River Avon. Sadly he would not live to see its construction. But Brunel was a restless man, always seeking new ideas, new creations, and he had an ambition to build the greatest railway ever known. In 1833 his wish was granted as he became chief engineer of the Great Western Railway that was to link London with Bristol, later extending to Exeter.

By the middle of the nineteenth century Brunel had been responsible for the construction of more than 1,600km (1,000 miles) of railway as his engineering achievements extended further across Britain. But it was during an earlier directors' meeting of the Great Western Railway (GWR) held in a London hotel that he was encouraged to change direction once more. Someone expressed their concerns about the great length of the proposed main line to Bristol, but Brunel was moved to respond: 'Why not make it longer and have a steamboat go from Bristol to New York, and call it the *Great Western*?' Laughter from everyone present greeted this apparently ridiculous suggestion, apart from one man, Brunel's closest Bristol-based colleague, Thomas Guppy. He and Brunel discussed the idea well into the night and, in June 1836, with Peter Maze, another pioneer of the GWR, as chairman and Captain Christopher Claxton, a nautical adviser, also on board as managing director, the Great Western Steamship Co. was formed.

The reason for the men's unwavering support for this seemingly ludicrous idea was Brunel's strong conviction that he could design a ship capable of carrying sufficient coal to traverse the Atlantic under the power of its own engines. 'It is a case of size and proportion,' he observed. A larger ship would carry more coal but would not use as much in proportion to that of a smaller ship. The North Atlantic had already been crossed by two vessels fitted with steam engines, but both had needed sail power for a substantial part of the voyage, hence the widespread distrust of Brunel's claims. In 1819 the 320-ton American sailing packet *Savannah*, equipped with an auxiliary steam engine and a pair of collapsible paddle wheels, sailed from Georgia, USA, to Liverpool, taking 663 hours, of which only eighty were made under her own steam. Nevertheless, the 34m-long vessel is regarded even to this day as the Atlantic's first steamship, and her sailing on the 22 May is celebrated each year in the United States as National Maritime Day. In many ways, though, a far greater achievement was the Atlantic crossing of the little Canadian ship *Royal William* in 1833.

Sailing from Nova Scotia, the 370-ton vessel, owned by the Quebec & Halifax Steam Navigation Co., steamed for almost three-quarters of her twenty-two-day voyage, setting off with some 300 tons of coal and a complement of seven passengers, including, notably, the Cunard brothers, Samuel and Joseph.

It was resolved that Brunel's first ship should be constructed to his own designs by William Patterson of Bristol under the close scrutiny of Guppy, Claxton and Brunel himself.

On 28 July 1836 work on the *Great Western* officially began as Patterson set up the sternpost, and over the following twelve months a workforce of 180 toiled over the 72m-long hull. Brunel may well be remembered as a 'Man of Iron', but the era of complete iron-built hulls had yet to arrive. Hence, the *Great Western* was constructed of oak, using traditional shipbuilding methods, although iron was used for strengthening, ensuring the ship would withstand the most potent Atlantic storms. Brunel had applied his tremendous knowledge and experience of bridge designs to the creation of the world's biggest ship.

The *Great Western* took shape as a four-masted schooner rig, complete with a pair of paddle wheels. Her wooden hull was sheathed in copper (apparently to protect the oak from a feared shipworm of the time, the Teredo navalis) and a series of iron bolts 4cm in diameter ran the whole length of her bottom frame. Brunel reported to the Great Western Steamship Co. that 'her floors overrun each other, they are firmly dowelled and bolted, first in pairs, then together, using 1½ inch [4cm] bolts, about 24 feet [7.32m] in length, driven in parallel rows.' Meanwhile, by the River Thames at Lambeth, the ship's propulsion machinery was under construction at the workshops of Maudslay, Sons & Field, regarded as the most respected marine engine builders of the day. Her 450hp side-lever engine was equipped with two cylinders that could operate both paddle wheels either simultaneously or one at a time, whilst her four boilers weighed 100 tons together, even when empty.

On 19 July 1837 the *Great Western* was floated out of her Bristol dock amid great pomp and ceremony. Captain Claxton smashed a demijohn of Madeira wine against the beautiful gilded figurehead of Neptune, escorted by two dolphins on her black bows. A gathering of company directors and dignitaries were treated to lunch in her main saloon before she voyaged under sail (the ship was equipped with sails as an auxiliary measure) to London where she would have her engines fitted and the saloon lavishly decorated. The 22.85m-long saloon was the largest room afloat and would be adorned with ornamental paintings and mirror lined recesses, the work of Edward Thomas Parris, the historical painter to Queen Victoria, creating, by all accounts, a most amazing public space. Sadly no photographs exist to show that the room's height was actually just 2.74m, far lower than artists' impressions implied.

A ship's maiden voyage is always a momentous event – from those Victorian years right up to the twenty-first century – but the departure of the *Great Western* on her first Atlantic crossing was surrounded not only by great excitement but by sheer jealousy. The birth of Brunel's Bristol company had immediately led to the formation of two rival companies, the British & American Steam Navigation Co. in London and the Transatlantic Steamship Co. in Liverpool. The London company, formed by Junius Smith, an American residing in England's capital, designed a ship to be called *Royal Victoria* (later completed as *British Queen*) that would be slightly larger than the *Great Western*, whilst the Liverpool rivals purchased a 1,150-ton vessel, the *Liverpool*, that was under construction for a certain Sir John Tobin. But Brunel was still winning the race to have a ship ready for transatlantic service, and both companies, in a frantic attempt to outpace him, chartered two smaller, newly built ships, originally intended for Anglo-Irish service. Both needed modification

Extending the line to New York. Brunel's first ship, *Great Western*, on her maiden transatlantic sailing in April 1838.

for the strenuous journey that lay ahead, and history shows that, with the *Liverpool* already out of the equation, a straight race ensued between the *Great Western* and Junius Smith's chartered vessel, the *Sirius*. But even before this began, Brunel's ship did her best to write herself off. Leaving the Thames on 31 March 1838, with Brunel, Claxton and Guppy on board, her forward boiler room and then her engine room became enveloped by fire. Thanks to the gallantry of her chief engineer, George Pearne, the fire was extinguished, but not before Brunel himself had been close to certain death, his fall of 6m from a collapsing ladder being broken, by sheer chance, by his colleague Claxton.

The *Great Western* was able to proceed on her way, having lost some twelve hours of time, but exaggerated rumours of her safety were now rife in Bristol, and fifty passengers booked on the maiden voyage asked for their money back. Consequently just seven passengers remained to make the transatlantic crossing. *Sirius*, meanwhile, had slipped away from London three days before the *Great Western*, carrying forty passengers and a crew of thirty-five, bound, so they hoped, for New York. This was indeed the case, but their tiny 705-ton ship had first to call at Cork to fill up with coal. When the *Great Western* set out from Bristol on Sunday 8 April, *Sirius* was already four days into her voyage.

So the race, a most unequal one, was on. Passengers on the *Great Western* sat back to enjoy top quality food and wine, completely oblivious of the rigours endured by the ship's sweating, toiling crew below them who were required to shovel coal at the rate of almost 50kg per minute to maintain their ship's speed. It was a similar nightmare for the crew of *Sirius*, with the added worry of watching their coal stock diminishing by the hour. On the morning of 23 April *Sirius* docked at New York after nineteen days at sea and with just 15 tons of coal left on board. That afternoon the *Great Western* steamed in, only a few hours after her rival and with 200 tons of coal still in her bunkers. She may have lost the race, but Brunel had won over his critics. Transatlantic crossings by large steamships were quite clearly the way forward from then on.

The *Great Western* returned to Bristol with sixty-eight passengers, and before her second voyage was complete Brunel was already planning a new ship bigger than her. But the little genius, in typical fashion, was not content with building a larger version of the same ship. Iron had been employed successfully as a method of reinforcing the *Great Western*, so it appeared perfectly logical that it should be totally used in the construction of his new ship. Once again Claxton, Guppy and Brunel joined forces to draw up plans and oversee construction at Patterson's yard. After four designs had been discarded, a fifth, for a vessel of 3,270 gross tons, much larger than any ship so far envisaged, was approved, and on 19 July 1839 the keel of SS *Great Britain* was laid.

Tenders for the new ship's paddle engines were received from three sources. Among these were Maudslay, Sons & Field, who had built the machinery for the *Great Western*, but, somewhat surprisingly, a young man under the name of Francis Humphrys was awarded the constructional work. Poor Mr Humphrys! His huge paddle wheels, upon which his whole business future was pinned, would have surely driven Brunel's new ship across the world but for the chance arrival in Bristol in May 1840 of the three-masted topsail schooner *Archimedes*. This ship was unique because she was the only sea-going vessel driven by a screw propeller, an invention of Sir Francis Pettit Smith, a sheep farmer from Kent, no less, who dabbled in model boats. The Great Western Steam Co. immediately chartered *Archimedes* for six months to enable Brunel to study and then re-adapt Pettit Smith's basic design for his new venture. By attaching a screw propeller to an iron-built hull and by adding a balanced rudder to control the ship's direction (sailing ships had always been steered by their sails) Brunel was producing the forerunner of the modern ocean liner.

As for Humphrys, a sensitive fellow, so devastated was he that work on his paddle engines had to be abandoned that, despite his young age, he passed away following a short illness.

In her Bristol dock, the hull of the *Great Britain* was steadily taking shape. Although constructed of iron, much emphasis was again placed by Brunel on its longitudinal strength, and ten iron girders ran along the entire length of the ship's bottom whilst to their upper flanges an iron deck was secured. Two longitudinal bulkheads divided the ship up to main deck level in addition to being divided into six compartments by five watertight transverse bulkheads. The engines and boilers would be placed in the large central portion consequently formed, while the side compartments would be coal bunkers. Without doubt, by creating this cellular form of construction, Brunel was very much influencing future ship designs, even to the present day. The engines of the *Great Britain* would consist of four inclined cylinders, whilst, after much deliberation, Brunel decided on a six-bladed single screw, just under 5m in diameter, with a 7.6m pitch and a normal shaft speed of 54rpm.

The *Great Britain* was named and launched – or, more precisely, floated out – on 19 July 1843 by the Prince Consort who had travelled by train from Paddington on Brunel's Great Western Railway line. Prior to the ceremony some 600 people sat down to a grand banquet, and then watched as the Prince cracked a bottle against her black bows, giving the signal for the dry dock sluices to be opened. The *Great Britain* was towed slowly out of the building dock, but with her length of 98.15m and width of 15.39m (she was at least 30m longer than any other ship so far built), she was both too long and too wide for Bristol's lock chambers. Much to the embarrassment and annoyance of Brunel, her exit was delayed by lengthy negotiations with the Bristol Dock authorities, and even after passing through one lock that had been temporarily widened, his new ship almost stuck fast in another.

At last, on 23 January 1845, the *Great Britain* sailed for the Thames to be fitted out. There she was visited by Queen Victoria and the Prince Consort, who could only admire her sixty-four staterooms and more than 1,000m of Brussels carpeting. She had been designed to accommodate 120 first class passengers, of whom twenty-six were in single-berth cabins, and 132 second class passengers, whilst her officers and crew totalled 130. Her fitting out complete, the great new ship sailed to Liverpool from where she left on her maiden voyage to New York on 26 August 1845. This splendidly successful voyage took fourteen days and twenty-one hours, and she returned in fifteen-and-a-half days. Brunel and his Great Western Steamship Co. now had two ships traversing the mighty North Atlantic.

Compared with the gigantic moving resorts that cruise our seas today, the two vessels were positively diminutive, but to the Victorians they were the wonders of their time. Brunel had proved his point, but when he began drafting his design for a ship more than 200m in length, it seemed even he had taken leave of his senses. But his idea, arguably not as ludicrous as extending his Great Western Railway to New York, was greatly influenced by an unfortunate accident to the *Great Britain*. The ship had proved utterly reliable on her first four round trips, but a voyage from Liverpool in September 1846 with a record 180 passengers on board ended in disaster. A few hours from port, panic-stricken passengers rushed from their cabins as the *Great Britain* shuddered and came to an abrupt halt. She had run aground on rocks in the middle of the night. Her crew had been using faulty charts and although her captain thought they were close to the Isle of Man, the morning light revealed they were, in fact, in Dundrum Bay, County Down, in Ireland.

Any other ship of that time would surely have broken in two, but the fact that the damage to the *Great Britain*'s hull was relatively small and that all on board were safely removed was testimony to her strengthened hull. Nevertheless, unfolding events resulted in Brunel's pride and joy remaining beached and open to the elements for many months, not returning to Liverpool until August 1847. The salvage operation financially crippled the Great Western Steamship Co., and it was forced to sell the *Great Western*, which continued in service under new ownership until 1856, and with no reliable source of income it had to put the *Great Britain* up for auction, resulting in the company being wound up.

This unhappy episode coincided with the beginning of a new wave of emigration to Australia that was to reach a new high following the discovery of gold in the 'promised land' in 1851. Under her new owners, Gibbs, Bright & Co., the *Great Britain* was repaired, her accommodation extended and she was placed on a service between Liverpool and Melbourne that she was to successfully maintain for some twenty years, carrying up to 630 passengers on each voyage. She had, of course, been designed for the North Atlantic, and the long journey to Australia exceeded her steaming range by some distance, making it necessary for her to stop more than once along the way to take on more bunker coal. With this in mind, Brunel was consulted by the Australian Mail Co. as to the optimum size of a ship for the route 'down under'. His decision was a vessel of 5-6,000 tons, requiring only one coal stop at the Cape of Good Hope.

However, this enquiry proved food for thought for Brunel. He mused:'What if I design a ship so large she could carry her own coals around the world?' A ship six times the size of anything else afloat would be the answer, and during 1852 he met up with his faithful friend Claxton and John Scott Russell, a Glaswegian who had overseen the building of four ships at Greenock in Scotland. Scott Russell had purchased a shipyard at Millwall, on the banks of the Thames, which had been founded by a William Fairburn. By this time Thomas Guppy, Brunel's other colleague of long-standing, had moved to Italy. Ideas for the 'Great Ship', as the project became known, were put to the Eastern Steam Navigation Co., a concern recently formed to operate passenger and mail routes to Australia via India and China. But in March 1852 the British Government offered the Australian mail contract to the Peninsular & Oriental Steam Navigation Co. (commonly abbreviated to P&O) that was, at the time, extending its own services to the Antipodes. Consequently, the directors of Eastern Steam were only too glad to talk further with Brunel about his new project, appointing him engineer.

Brunel estimated the cost of his new ship to be around £500,000, but the actual figure tendered by Scott Russell was only £377,200, made up as follows: hull £275,200, screw engines and boilers £60,000 and paddle engines and boilers £42,000. From this information it was clear that Brunel was reckoning on fitting both a propeller and paddle wheels to drive the ship. Although a proviso in the contract allowed for continuing revision of calculations and drawings, Brunel must surely have been surprised at Scott Russell's comparatively low tender.

In designing the ship's hull, Brunel continued with and further developed the cellular system employed in the construction of the *Great Britain*. The iron deck would become an inner watertight skin extending across the bottom of the ship and even up the sides to lower deck level, almost 12m above the keel plate. The space between the two skins of the double hull would be almost 1m, and the joints between them divided the space into a series of cells. The main deck would be similarly constructed, creating an extremely strong hull, whilst iron watertight bulkheads would divide the ship into ten 18m compartments. As on the *Great Britain*, two longitudinal bulkheads, 11m apart, would be incorporated, extending the whole length of the two engine rooms and boiler rooms up to the loadline. The double hull itself would be constructed of 19mm wrought iron in 0.86m plates attached to ribs every 1.8m.

So where could a ship of such magnitude be built? The area of Scott Russell's Millwall yard was completely inadequate, but, by luck, an adjoining yard, owned by another Scottish shipbuilder David Napier, until his retirement twelve months earlier, was standing vacant. The entire ship could be assembled in the Napier Yard, allowing Scott Russell's premises to deal with the manufacturing work. But that was not all: how would the ship be built and launched? There were three options: the first, to build the huge construction in a dock, was immediately dismissed. The second was to build on an end-on slipway (that remained the most popular system until halfway through the twentieth century), but Brunel's calculations showed that, based upon the ideal gradient of the ship, the top of her bows would be some 30m in the air, causing great inconvenience to the ship workers. So the third option, to build the ship parallel with the river bank and launch her sideways, was deemed the favoured choice.

Everything about the *Great Eastern*, as the ship would eventually be called (although for a while she was referred to as 'Leviathan') was full of superlatives. As well as her 208m (693ft) length, her beam, including paddle boxes, measured 36m (120ft) and she had a gross tonnage of 18,915, a statistic that would not be exceeded by any other ship until the advent of the twentieth century. Like the *Great Britain*, she was equipped with sails which were affixed to six masts, said to be named after the six days of the working week (Monday being the foremast, and Saturday the spanker mast) and she could carry 12,000 tons of coal, ensuring she could journey to Australia without the need to refuel. The greatest superlative, though, was surely her ability to accommodate no fewer than 4,000 passengers, an incredible number which today would place her close to the top of the list of the world's highest capacity passenger ships.

The keel was laid on 1 May 1854, and historical records show that it took some 2,000 workers almost four years to complete the ship. In truth, however, this period of time was fraught with problems. Unbeknown to Brunel, Scott Russell was in financial difficulties and his 'competitive' tender, which was a million nautical miles below the true constructional cost, only managed to exacerbate his problems. Both he and Brunel were men of strong character, and Scott Russell's continual avoidance of Brunel's requests for information led to much animosity between them. Constructional work on the giant ship was erratic to say the least, and actually ceased for three months during 1856 as Scott Russell came ever closer to bankruptcy. Eventually Brunel could take no more and announced that only if the Eastern Steamship Navigation Co. were allowed to move in on the act would the ship ever be finished. Only after many weeks of argument with Scott Russell's creditors was this agreed, and the company was shocked to find the ship only a quarter complete, despite Russell having been paid almost £300,000. The stress of this sequence of events was already taking its toll on Brunel's health, and just when he thought his problems were behind him, he found himself faced with further setbacks.

The 'little giant' had produced countless calculations to ensure the smooth launch of his mammoth ship. In an attempt to recoup at least some of the additional construction costs, the ship's owners sold thousands of tickets to sightseers and Brunel was especially keen to impress them. The launch had already been significantly delayed when, on the morning of 3 November 1857, the first attempt to move the ship was made. The *Great Eastern* was supported by two cradles, each about 37m wide, in such a way that her bow and stern projected beyond them and a length of almost 34m was unsupported in-between. The cradles were attached to heavy chains that were wound around huge checking drums. At the time of launch, gearing handles attached to the drums were turned to pay out the chains, and then winches mounted on the shoreline and on moored barges were started up. As the ship refused to budge, Brunel demanded that hydraulic presses be brought into play, but the ship moved no more than a metre.

A further attempt was made that afternoon, but the result was little better and the crowd of onlookers who had come to witness a memorable moment became restless and trooped home disappointed. The most protracted ship launch in history continued for ninety days until, at daybreak on 31 January 1858, an exceptionally high tide accepted the *Great Eastern* into the water. The overall building and launching costs ruined the Eastern Steam Navigation Co., and the ship was purchased by the Great Ship Co. who officially registered her name. Brunel's third ship had been conceived for Australian service but never fulfilled this purpose, being fitted out for transatlantic service instead, sailing between Southampton and New York. Brunel failed to fulfil his dream. Just before the *Great Eastern* set sail for a promotional voyage to England's south coast ports, he posed for a picture by the ship's funnel. Shortly afterwards Brunel was paralysed by a stroke. As the *Great Eastern* steamed at full speed in the English Channel, a huge explosion launched her forward funnel into the air. Scalding hot water poured into the engine causing fatal injuries to five stokers. Within six days of hearing the news of this disaster Brunel passed away.

As testimony to Brunel's brilliant engineering ability, the *Great Eastern* simply continued to steam on despite the damage caused by the explosion and, following repairs, made her eleven-day maiden voyage on 17 June 1860. On board were just thirty-five paying passengers, eight company non-paying passengers and 418 crew! Regrettably, although response to the new 'wonder ship' improved a little, her passenger accommodation was never filled to capacity and she was chartered by the Telegraphic Construction Co. to lay the first communication cable beneath the Atlantic. With funnel No.4, some boilers and most of her cabins and public areas removed to make way for cable tanks, the *Great Eastern* began her new role in 1865. At last she enjoyed some success in her otherwise troubled career, laying some 4,200km of transatlantic telegraph cable.

The *Great Eastern* was eventually broken up at Rock Ferry on the River Mersey. Dismantling started in 1889, but so heavily reinforced was she that her breakers spent eighteen months taking her apart. To this day a small part of Brunel's biggest ship remains in public view, her top mast having been purchased for use as a flag pole at the 'Kop' end of Liverpool Football Club's Anfield ground.

As for the *Great Britain*, her story is considerably happier and she now sits proudly in Bristol Dry Dock where she was built, a fitting monument to Brunel. She had continued her successful employment as an emigrant ship on the Australian run for a quarter of a century (save for a three-year stint as a troopship during the Crimean War and the Indian Mutiny), but in 1882 she was converted into a sailing ship for transporting bulk coal. Badly damaged in 1886, she was considered to be beyond repair and was used as a storage hulk for wool and then coal in the Falkland Islands until 1937 when she was scuttled and abandoned. Her days were surely over, but once again Lady Luck smiled upon her when, in the 1960s, a growing interest in the subject of preservation led to a huge fund-raising campaign to save her. In 1970 she was re-floated onto a pontoon and towed all the way back to her birthplace at Bristol where today the *Great Britain* and the adjacent Dockyard Museum are open to the public. One of the many fascinating artefacts on view is the bell of Brunel's first ship, the *Great Western*, which had continued in transatlantic service until 1846, and, after serving as a troopship in the Crimean War alongside the *Great Britain*, was broken up at Vauxhall on the Thames.

During the years embracing Brunel's engineering career and immediately after his untimely death, a number of businesses that would later grow into massively influential maritime organisations were still very much in their infancy. As we have seen, Canadian shipping magnate Samuel Cunard and his brother Joseph crossed the Atlantic to England in 1833, and so convinced was Samuel that a regular transatlantic service was the way forward, he collaborated with Scottish engineer Robert Napier and businessmen James Donaldson, Sir George Burns and David MacIver to form the British & North American Royal Mail Steam Packet Co. In 1839 the company was awarded the Northern Atlantic mail contract and within a year was operating a pair of paddle steamers, *Unicorn* and *Britannia*, out of Liverpool. Both the vessels were only a fraction of the size of Brunel's ships, but *Britannia* will always be regarded as the first holder of the Blue Riband, an accolade awarded for the fastest transatlantic crossing, that would strongly influence the size and speed of liners of the future.

Samuel Cunard died in 1865, but his two sons, Edward and William, continued to play vital roles in the company which became known as The Cunard Steam Ship Co. Ltd, set up in 1878. With the valuable mail contract firmly in its grasp, the Cunard company was proud to be allowed to prefix all its ships' names with the three letters RMS (Royal Mail Ship). Whilst Brunel had his ships built at Bristol and London, Cunard's vessels originated from Scotland. *Britannia* and two subsequent sisters were the work of one of Samuel Cunard's business partners, Robert Napier (no relation, apparently, to David Napier), under whose guidance two brothers, James and George Thomson, were apprenticed. The young engineers set up a marine engine building business in an area of Glasgow in 1847, and their reputation for reliability enabled them to open a small shipyard nearby. The Clyde Bank Iron Shipyard was later left to George's son, James Rodger Thomson, and in 1871 some thirty-two acres of farmland was acquired for a new, more expansive yard at Barns of Clyde, conveniently positioned opposite the mouth of the River Cart, into which the huge passenger liners of latter years would be launched. The shipyard took on the Clydebank name that would later be afforded to a brand new town that grew up around the yard. By the end of the nineteenth century the Clydebank company had become a highly profitable concern and was taken over by Sheffield-based forgemasters and armour plate makers, John Brown & Co., who were eager to involve themselves in shipbuilding. From then on the names of Cunard and John Brown became almost synonymous as the shipping company ordered its largest and most famous liners from the Scottish yard.

British shipbuilding was leading the world, but from across the Irish Sea, John Brown & Co. found serious competition from a Belfast yard. Harland & Wolff was formed in 1861 by Edward James Harland, an English design engineer, and Hamburg-born draughtsman Gustav Wilhelm Wolff. Harland had worked as general manager of a small shipyard on Queen's Island before buying the business, appointing his German colleague as partner. Initially, thanks to Wolff's connections, the company built small steamers for the newly formed Hamburg America Line (Hamburg Amerikanische Packetfahrt Actien Gesellschaft, or HAPAG for short), but it was soon to follow a similar path to that of its Glasgow counterpart, forging a long-term link with another growing North Atlantic shipping company, the White Star Line. This concern, founded in Liverpool in 1845 by John Pilkington and Henry Wilson, had originally focused on the growing Australian gold mine trade by developing a fleet of clipper ships for the service. Within twenty years, however, following an inadvisable merger, their business was suffering from financial problems. Cue the arrival on the scene of a certain Thomas Ismay, a young Liverpool ship-owner, to save the day. He took on board both Edward Harland and Gustav Wolff as shareholders, and in 1871 their shipyard delivered the first White Star liner under

White Star Line's first *Oceanic* of 1871 had the luxuries of promenade decks and bathtubs for her passengers.

the new partnership. *Oceanic* opened the doors to a new era of interior design, offering spacious passenger accommodation adorned with ornamental fittings, ideas that Harland & Wolff's chief designer William James Pirrie had taken from principal hotels ashore. Pirrie later became chairman of Harland & Wolff following Edward Harland's death in 1894.

By this time Cunard had responded to White Star's *Oceanic* with the launch of their 7,391-gross-ton *Servia*, the first liner to feature electric incandescent lamps throughout her interior and the first Cunarder constructed with a steel hull. The race for the Blue Riband was now really on, with serious competition coming from the opposite side of the 'pond'.

American ship-owners Inman Line had begun operating steamers in the 1850s, but it was their superb pair of 10,500-ton liners, *City of New York* and *City of Paris*, introduced around 1890, that caught British ship-owners off their guard, stealing the coveted award from Cunard. The duo were also the first vessels to exceed 10,000 tons since the *Great Eastern*, and with Brunel's ship now employed in cable-laying, were statistically the largest passenger liners afloat. But yet again Cunard responded with two even larger ships. *Campania* and *Lucania* both measured 12,950 gross tons, were 204m long and could carry 2,000 passengers each. Built by the Fairfield Shipbuilding & Engineering Co. at Govan, Scotland, they incorporated a number of 'firsts' into their design, including private bathroom facilities in the cabins and purpose-built single-berth cabins, not to mention that they were the first liners to be driven by two propellers.

The style of their interiors was predominantly Art Nouveau, and they represented, in many people's opinion, the ultimate in Victorian opulence. Below deck, two five-cylinder triple-expansion engines gave a service speed of 22 knots, enabling Cunard to snatch back the Blue Riband from the Americans.

As time moved relentlessly on, the new century drew ever closer – a century that would produce ships of a size and quality unimaginable in the Victorian age. But, despite the giant steps thus far achieved in shipbuilding and design, the size, in terms of tonnage, of Brunel's biggest iron ship, the *Great Eastern*, had yet to be surpassed.

CHAPTER TWO

FOUR-FUNNELLED FERVOUR

Their look spelled out power and speed; they were prestigious national symbols, combining engineering accomplishment with a handsome image. The four-funnelled liners had arrived and would stay supreme in history forever. Yet it was not Great Britain or even America that was leading the way, but Germany. At a time when British shipyards were building 70 per cent of the world's shipping, the Germans, eager to be recognised as a leading sea power and encouraged by Kaiser Wilhelm ll himself, planned a new breed of North Atlantic greyhounds that would be constructed in their own shipyards, using skills and even workers borrowed from Britain. With more than one eye on the Blue Riband, Germany ensured that the ships they built would be fast and that they would seize their fair share of an expanding transatlantic market with their interior grandeur.

North German (Norddeutscher) Lloyd and the Hamburg America Line had both represented Germany's North Atlantic presence for a number of years, and it was the former company that took the initial plunge by instructing Vulcan Shipyards at Stettin (now known as the Polish port of Szezecin) to construct a 200m-long ship that would be the world's first four-funnelled ocean liner. Launched on 4 May 1897 by the Kaiser, she was named *Kaiser Wilhelm der Grosse* (Kaiser William the Great) after his grandfather, the first emperor of the new German Empire.

Over the following decades a voyage on a four-funnelled passenger ship was regarded as the ultimate in sea travel, yet, like Germany's first major transatlantic liner, most of these vessels did not really need that many smoke stacks. Indeed, *Kaiser Wilhelm der Grosse* had only two uptake shafts from her boiler rooms that branched in two to fit her four funnels, hence she appeared to have two sets of two funnels rather than four evenly spaced stacks. Measuring 14,349 gross tons, she could accommodate 1,506 passengers (1,074 of whom were third, or steerage, class) and was powered by the well-proven system of triple-expansion reciprocating engines driving two propellers and giving a service speed of 22.5 knots. *Kaiser Wilhelm der Grosse* set off from Bremerhaven on her maiden voyage on 19 September 1897, and in the following November fulfilled Germany's prime aspirations by breaking the eastbound North Atlantic crossing record between Sandy Hook and the Needles, on the Isle of Wight, averaging 22.35 knots. Four months later the westbound Blue Riband was hers as she snatched the invisible trophy from Cunard's *Laconia*. Britain was shaken and would soon be stirred, yet would not take back the coveted prize for another decade. The Germans, no doubt buoyed by their initial success, went on to produce four more four-stackers – three for North German Lloyd and one for Hamburg America – as if to rub salt into Britain's wounds. At least the British had one trump card up their sleeve, even though the vessel in question could only manage half their number of funnels.

Thomas H. Ismay's philosophy that luxury should take precedence over speed meant that his latest and biggest White Star liner would never keep pace with her German rivals but could overtake them in terms of size. Furthermore, although her gross tonnage of 17,272 would still not top that of Brunel's *Great Eastern*, she would, at least, out-measure her by about 4m, having a length of some 215m. Ismay, as usual, turned to Harland & Wolff in Belfast, this time with instructions to W.J. Pirrie, supervisor of the design work, that the shipyard should build him 'nothing but the very finest'.

Oceanic, as the new ship would be named, certainly transpired to be a good-looking vessel, with two tall and slender funnels and three even taller masts. Named after her predecessor of 1871, she needed 1,500 workers to build her and was launched on 14

The first of the inspirational German four-stackers, *Kaiser Wilhelm der Grosse*.

January 1899. Due to her length, the usual method of launching, involving the use of large pieces of timber as struts between the fixed and sliding ways, that would be knocked away to enable the ship's cradle to glide towards the waters, was replaced by the positioning of a huge steel trigger activated by a hydraulic cylinder. As with all White Star liners before her, *Oceanic* had no godmother to name her, but her launching, despite being a comparatively low-profile affair, caused the gathering of a reported 50,000 onlookers, crammed together on the adjacent river bank where they could view the event for free. Seating for some 5,000 guests had been erected much closer to the slipway, but the price of 10s per ticket dissuaded many people from enjoying the privilege.

Oceanic was an extremely comfortable ship for her day, with a sedate service speed of 19.5 knots. She would sail between Liverpool and New York with a brief call at Queenstown (now known as Cobh) in southern Ireland, a voyage that would take a week. Like *Kaiser Wilhelm der Grosse*, she was equipped with triple-expansion reciprocating engines, but generated about 5,000hp less than her German counterpart. Setting sail on her maiden transatlantic crossing on 6 September 1899, she had room for 1,710 passengers, 410 of which were first class, accommodated across four decks amidships, the highlight of whose facilities was undoubtedly a white and gold-decorated dining room topped by a central glass dome. The 300 second class passengers could enjoy music rooms, a fine library and a gentleman's smoking room, whilst the 1,000 passengers booked into steerage class, the 'cattle-truck' of liner accommodation at that time, could experience what was described as being 'in advance of anything previously seen on ocean liners'. Despite this glowing description, however, male and female passengers were segregated in separate dormitories, an arrangement that would be employed on Antipodean migrant ships a number of years later.

Costing around £1 million to build, *Oceanic* was soon dubbed 'Queen of the Ocean' and operated successfully until the First World War, and was one of the rescue ships involved in retrieving bodies from the sinking *Titanic* in 1912. She met her own fate in 1914 during her first deployment for the British Royal Navy, running aground near the Shetlands and being declared a total loss. Ironically, *Kaiser Wilhelm der Grosse*, *Oceanic's* principal rival, also ended her life in 1914, sunk in action and having the dubious claim to fame of being the first converted passenger liner to be lost in the war.

Oceanic was the world's biggest liner until 1901 when her owners launched the 21,035-gross-ton *Celtic*, the first ship to exceed 20,000 gross tons and the first vessel to surpass the *Great Eastern's* gross tonnage. Moreover, she was the first White Star liner to be built since the death of founder Thomas Ismay on 23 November 1899. Meanwhile the Germans continued to concentrate on their powerful-looking four-stackers, with the launching in 1900 of SS *Deutschland* for Hamburg America. Of similar appearance to *Kaiser Wilhelm der Grosse*, she measured 16,502 gross tons, had a length of 208m and a speed of 22 knots, and she soon took over the prize of the Blue Riband from North German Lloyd's ship, traversing the Atlantic in a little more than five days. With her rather overstated, gilded interior she was nevertheless an extremely popular ship from the outset, until, that is, her powerful quadruple-expansion engines began to vibrate.

The constant rattling and excessive noise as the ship ploughed her Atlantic furrows soon turned her customers against her, and she became dubbed 'The Atlantic Cocktail'. In 1910 her owners finally put an end to these problems by converting her for cruising, and she was later deployed as an emigrant carrier until sold for scrap in 1925.

The four-funnelled fervour continued with the the completion of a second liner for North German Lloyd, the SS *Konprinz Wilhelm*. Launched on 30 March 1901 she was smaller than *Deutschland*, having a gross tonnage of 14,908 and a length of 202.17m, but she was fitted with two ultra-powerful four-cylinder quadruple-expansion steam engines, thus making her especially speedy and on her Bremerhaven-Southampton-Cherbourg-New York run she broke the record for the final leg of the voyage, crossing from France to America in five days eleven hours and fifty-seven minutes at an average speed of 23.09 knots. During the First World War her potent machinery influenced her deployment as an auxiliary cruiser, but in 1917 she was captured by the Americans, following the United States' declaration of war on Germany, and was commissioned into the US Navy as *Von Steuben* in honour of Baron Friedrick Wilhelm von Steuben, German hero of the American Revolution. She remained in American hands until her demolition in 1923.

The fourth of Germany's four-stacker quintet proved to be the biggest (by all of 1 gross ton), measuring 19,361 gross tons and 215.9m in length. SS *Kaiser Wilhelm ll*, the second ship to bear this name, was launched at the A.G. Vulcan Yard on 12 April 1902, and began operating regularly on the North Atlantic on 14 April the following year. Although

An abundance of panelling adorns the tourist class lounge of the liner *Deutschland*.

clearly larger than *Oceanic*, she was eclipsed as the world's largest liner by White Star's *Celtic*, but, with her quadruple-expansion steam engines powering to 23 knots, she won the Blue Riband for the fastest eastbound crossing in 1904. The carriage of third class passengers was still proving an important money earner, and of her 1,888 passengers a considerable proportion were accommodated in her lowest class. As soon as war reared its ominous head, she was captured by the Americans, renamed *Agamemnon* and employed as a troop transporter. Laid up in the 1920s under the name of *Monticelli*, she remained out of service until 1940 when she was scrapped. North German Lloyd later rounded off the quintet with a close sister to *Kaiser Wilhelm II*. The 19,360-gross-ton SS *Kronprinzessin Cecilie* (Crown Princess Cecilia), yet another product of Stettin, could accommodate 1,741 passengers and sailed on the Bremerhaven–Southampton–Cherbourg–New York route from 1906 until the outbreak of war. Torpedoed in October 1918, she survived, only to be laid up, like her sister, until 1940 when she was sent to the breakers.

So Germany had built its last four-funnelled liner but had opened the door to a new era, an era that announced to ship-owners and shipbuilders alike that 'the more funnels an ocean liner had, the better'. Sea travellers would fairly flock to voyage on these imposing multi-funnelled vessels, regarding the four-stacker as the ultimate example of safety as well as power and might. But for the moment at least the title of the world's biggest liner was in the hands of a ship that could muster just two funnels, White Star Lines' 23,876-gross-ton *Baltic*. She entered service on 29 June 1904 and was capable of carrying no fewer than 2,875 passengers, in response, no doubt, to the ever-buoyant North American emigration market. With the earlier *Celtic* and *Adriatic* she was the third of 'The Big Four', as White Star's latest quartet was known.

But the Germans refused to lie down, Hamburg America Line unveiling their 22,225-gross-ton two-funnelled *Amerika* in October 1905. Led by a most unremitting man named Albert Ballin, whose high demands had produced a supremely well-run company, Hamburg America had contracted Harland & Wolff to build their biggest and most sumptuously appointed ship yet. The Belfast yard had habitually been linked with the White Star Line, its construction techniques having long been admired by the Germans. But English-speaking Ballin was to forge a further valued connection: French architect Charles Frederick Mewès.

In collaboration with Cesar Ritz, Mewès had designed the Ritz Hotel in Paris and then created the Palm Court and Grill in London within the refurbished interior of Carlton House Hotel. Mewès engaged a young English architect, Arthur Davis, with whom he had studied in Paris, as business partner for his London projects, designing the London Ritz Hotel and later the Royal Automobile Club in Pall Mall. But by this time Ballin had appointed Mewès as an interior designer to the *Amerika* who was able to offer her passengers many innovations, including a winter garden, electrical medicinal baths and, naturally, a Ritz-Carlton restaurant. The liner sailed on her maiden voyage from Hamburg on 11 October 1905, calling at Dover and Cherbourg en route for New York, and spent nine years on the Atlantic run. Mewès was now Hamburg America's resident interior designer and the Mewès-Davis influence would be evident on transatlantic liners for years to come.

Yet the beautifully decorated *Amerika* that was even equipped with electric lifts for her passengers was soon upstaged by a further two-funnelled German liner, this time home-produced at Stettin – a ship that would go on to acquire a well-known name in British maritime history. Hamburg America's SS *Kaiserin Auguste Victoria* was to have been named *Europa* as a sister to *Amerika*, which had been launched only a few days before. However, at the last moment, the German Empress Victoria, who performed the ship's christening, gave permission for the vessel to be named after her. Measuring 24,581 gross tons, the ship was, for approximately a year, the world's largest liner. Able to accommodate 1,897 passengers, she was distinctive by being fitted with three masts in addition to her two yellow funnels, whilst her interior decor very much bore the Mewès stamp. Taken over by the Americans by the end of the First World War, she operated as a troopship until being refitted and transferred to Canadian Pacific ownership and renamed *Empress of Scotland*, the first vessel with that celebrated name. For just over eight years she plied between Southampton, Cherbourg and Quebec, until she was broken up at Blyth in 1930.

During those early years of the twentieth century ships were growing bigger by the year amid a private battle between the principal German passenger liner companies and the White Star Line. So what was the Cunard Steamship Co. doing about this, I hear you ask, as, after all, it had not owned a ship that could call itself the world's largest liner since the *Lucania* in 1893? Moreover, the company's concerns were increased by the sudden intervention into the liner trade by American tycoon John Pierpont Morgan, the world's richest man, who formed the International Mercantile Marine (IMM), buying out six British passenger ship companies, including Ismay's White Star Line, no less, for an incredible $25 million – in cash! Surely with seemingly inexhaustible capital to hand he would next move for Cunard? This opinion was also held by Cunard chairman Lord Inverclyde, who wasted no time in obtaining Government agreement that his company's Articles of Association could be amended to prevent such a takeover. Furthermore, it was widely felt that the transfer of White Star to an American-owned venture might leave the British Admiralty bereft of liners suitable for conversion into armed merchant cruisers should war ensue. So Lord Inverclyde re-approached the Balfour Government, requesting a loan for the construction of the biggest and fastest liners ever built; two ships that, if called to war, would provide valuable support to the British Navy. In 1903 an agreement was signed for a loan of £2.6 million, plus an additional twenty-year subsidy of £150,000 per annum, to cover the cost of ensuring that the liners would be capable of being armed and enabling the Admiralty to have first claim to the ships for war duties.

Contracts for the construction of the two liners, which would be named *Lusitania* and *Mauretania*, were awarded to John Brown & Co. Ltd, Clydebank, and the Tyneside firm of Swan Hunter & Wigham Richardson, respectively. They would each have a speed of at least 24.5 knots, swift enough to see off any competition from the German four-stackers.

But that was not all. Rather than persisting with well-proven steam reciprocating engines, Cunard opted to install steam turbines, the latest innovation in marine technology, evolved by the gifted Irish-born gentleman engineer Charles Algenon Parsons.

Born in 1854, one of six sons of the Third Earl of Rosse, himself an engineer, Parsons developed his first steam turbine engine at the age of thirty, whilst working for a Gateshead company. This was used primarily for electric lighting in ships, but it was soon apparent that the steam turbine had great potential throughout the electrical industry, and before long Parson's turbo-generators were being supplied to power stations. But Parsons decided to move on independently and formed a company, C.A. Parsons & Co., with some friends at Heaton, Newcastle. He began work on adapting his steam turbines to propel ships and designed and built his own steam yacht, the *Turbinia*, which would soon justify his theories. Just 31.62m long, the craft was equipped with three parallel-flow steam turbines, each driving a separate shaft on which three propellers were fitted. She responded spectacularly, racing to 34.5 knots on her North Sea trials and displaying her abilities at Queen Victoria's Diamond Jubilee Fleet Review at Spithead in 1897. The maritime world was won over.

HMS *Viper* became the world's first steam turbine-powered warship in 1900, and a year later the Firth of Clyde steamer *King Edward* entered service as the first turbine-propelled passenger ship. In 1905 a pair of American-owned Allan Line ships, *Victorian* and *Virginian*, employed the steam turbine to cross the Atlantic, and in that same year the Cunard company introduced two brand new sister ships, *Carmania* and *Caronia*, the former equipped with Parsons' turbines, the other with customary quadruple-expansion engines. Cunard's initial assessment found *Carmania* to be not only faster but more economical and efficient. So their question was answered: both *Lusitania* and *Mauretania* would be fitted with four massive Parsons' steam turbines. What is more, they would be Cunard's first four-stackers, the first passenger liners to exceed 30,000 gross tons and their machinery would be capable of generating 75 per cent more power than steam reciprocating engines of the same size. Surely Germany's ship-owners would finally meet their match.

The two 'superliners' took shape almost simultaneously, although the construction of *Lusitania* was always slightly ahead her sister. Laid down at Clydebank in May 1905, her hull incorporated a cellular double bottom and her coal bunkers were positioned along the sides of the boiler rooms offering protection for her machinery areas placed below the waterline. It was thought that her 68,000IHP turbines would otherwise be particularly vulnerable in the event of a collision or wartime enemy action. *Lusitania*'s design would enable her to carry twelve 6in guns should she be converted into an armed merchant cruiser.

To the south-east of Clydebank, careful preparations had been made for *Mauretania*'s construction on the banks of the River Tyne. Firstly, hull tests were run by employing a specially built electrically driven model of the liner at 1/16th scale. Tests took two years to complete before the engineers were happy to proceed. But there was a geographical problem to be overcome as well, for *Mauretania* would be longer than the width of the river at the location of her construction. By luck there was a conveniently placed bend in the river that could be utilised, and the ship's slipway was readied with some 16,000 piles being driven into the riverbank. Three rows of blocks were positioned on the berth floor, the centre block, the longest, being used to support *Mauretania*'s keel. Finally, to protect her Tyneside shipyard workers from severe winter conditions, a glass and iron roof, more than 228m long, was erected, complete with lighting and seven overhead electric cranes.

At both yards *Lusitania* and *Mauretania* began to resemble monster skeletons as steel frames were systematically attached, port and starboard, to the double bottoms. We shall study in more depth this method of shipbuilding in the next chapter, which emanated from the days of timber-built ships and continued until at least halfway through the twentieth century. Construction of the two liners progressed with the fabrication of the ships' outer skins by attaching shell plating to the completed framework. Plates almost 3cm thick were used, 15m in length and weighing 5 tons, which would then be riveted. The hull and superstructure of each vessel required four million rivets to assemble.

Lusitania was named by Lady Inverclyde, wife of the Cunard chairman, on 7 June 1906, sending some 16,000 tons of metal down the ways in two minutes and five seconds before it was checked by the same drag chains that were used in the launch of Brunel's *Great Eastern*. She was just under 2m shorter than her Clyde-built sister (5ft to be exact), but until *Mauretania*'s completion she would, at 30,996 gross tons, be the biggest ship afloat. Over the coming months the most noticeable change to her external profile was the positioning of her four mighty red- and black-painted funnels, which, incidentally, were all genuine outtakes (not one dummy among them). Indeed, there had been talk of a three-funnelled design for both vessels, but Cunard thankfully elected for four evenly

Following her successful launch, the mighty *Mauretania* is manoeuvred to her fitting out berth.

spaced stacks, expertly slanted to enhance the ships' graceful lines. *Mauretania* emerged from the shelter of her slipway's glass and iron roof on 20 September 1906, christened by the Duchess of Roxburghe, the 240.8m hull successfully avoiding the opposite riverbank in a perfectly orchestrated launch. Like her sister, she would be fitted out to accommodate 2,165 passengers (563 in first, 464 in second and 1,138 in third class), with the first class cabins and public rooms positioned amidships, away from the noise of her propellers and those areas most susceptible to movement in heavy seas.

Lusitania and *Mauretania* were deemed to be sisters, but their respective designers and builders had been afforded ample scope by Cunard to create their own individual ships, thus generating a unique atmosphere onboard each vessel. Externally, *Lusitania* was bereft of *Mauretania*'s distinctive top-deck ventilator cowls, whilst there was a variation in the design of each ship's forward superstructure. Below the waterline, *Mauretania*'s turbines were fitted with more blades than her sister (which would later give her a speed advantage). Internally, their decor was clearly understated compared with the lavishly ornate public rooms of their German competitors. Lord Inverclyde had personally appointed two eminent onshore architects, who had never before designed shipboard facilities, to produce surroundings in which British voyagers, in particular, would feel at home.

For *Lusitania*, James Miller, a Scottish architect, chose light woods trimmed in gilt and airy glass domes. The first class dining room was typically Louis XVI-style, whilst second class passengers fared almost as well in similarly decorated rooms. In contrast, Harold A. Peto was a designer of English country houses and carried over this style to *Mauretania*, with an abundance of darker woods and panelling reminiscent of gentlemen's clubs. In fact there were some twenty-eight different types of wood throughout the ship's interiors, especially abundant in her first class areas, with mahogany in the Louise XVI-style lounge, walnut in the Italian Renaissance-style Smoking Room and weathered Austrian oak in the imposing triple-deck dining room. Innovative features on the liner included a Verandah Café situated on Boat Deck and a series of electric lifts adjacent to her main staircase.

Mauretania measured 31,198 gross tons on completion and was to gain a fine reputation from the outset, a reputation that would grow to greater heights throughout her commercial life. As keen ship enthusiasts will know, she has been held in iconic status alongside such prestigious liners as *Queen Mary*, *Normandie* and the *QE2* ever since, whilst, sadly, her sister *Lusitania*, her life so cruelly cut short in 1915, was never allowed a chance. However, it was *Lusitania* that first proved her worth to Cunard, regaining the Blue Riband for Britain, not on her maiden voyage of 7 September 1907 but on her second voyage that began at Liverpool on 5 October, making the crossing from Daunt's Rock near Queenstown to Sandy Hook in four days, nineteen hours and fifty-two minutes at an average speed of 23.99 knots. Not content with this, she completed her homeward passage in record time, ensuring that British prestige had at last returned to the North Atlantic. *Mauretania* was delivered by John Brown & Co. and immediately showed on her acceptance trials that she would be even faster than her sister. She departed from

High and dry. *Mauretania* receives a well-deserved refit during her later years.

Liverpool on her maiden voyage on 16 November 1907 and broke *Lusitania*'s eastbound record on her return journey. Both vessels incurred problems with their propellers during their first year, despite their impressive performances, but once *Mauretania* received a new set in 1909 she broke the westbound record at a remarkable average speed of 26.06 knots, and continued to hold the Blue Riband for both directions for the next twenty years.

Both ships continued to operate superbly, but on declaration of war *Mauretania* was refitted as a troop transport while *Lusitania* became part of Cunard's reduced wartime transatlantic service. Her tragic loss on 5 May 1915 off the Old Head of Kinsale, Ireland, has been documented many times before, so I shall dwell no longer on one of the worst disasters in maritime history in which 1,198 lives were taken by a single torpedo fired from a German U-20 submarine. Today she lies in almost 90m of water on her starboard side, deteriorating so rapidly that only her bows are intact. *Mauretania*, by contrast, survived the First World War to prosper as a liner, then as a cruise ship, until making her mournful way to the breaker's yard at Rosyth in 1935.

Both Cunard's famous Edwardian sisters were committed to the annals of history, ironically never to have seen conversion into armed merchant cruisers, the principal condition of the British Government's loan to build them.

As *Lusitania* and *Mauretania* were making their triumphant entries onto the Atlantic scene in 1907, that other British giant, White Star Line, was spurred on to even greater things. J. Bruce Ismay, the son of the company's founder, had taken over as chairman following his father's death and was appointed as managing director and chairman of IMM, White Star's new American owners, by millionaire boss J.P. Morgan in 1904.

Edwardian elegance. The first class Smoking Room of *RMS Mauretania*.

BUILDING THE BIGGEST

In that decisive year of 1907, on a balmy July evening, Ismay and his wife were being entertained by the Right Hon. Lord Pirrie, chairman of Harland & Wolff, at his copious Victorian mansion in London's Belgrave Square. Ismay commented on how Cunard Line was enjoying all the plaudits for taking back the Blue Riband with their pair of express liners and how he wished to challenge them, not in terms of speed, but by building a ship brimming with sheer luxury. She would be the finest passenger liner the world had ever seen. But, just the one....? Well, maybe two, to match Cunard's duo, or even, as the wine began to flow, three! After all, money was no object. They would be named *Olympic*, *Titanic* and *Gigantic*, perfectly descriptive of their intended size.

The two men retired to the smoking room to enthusiastically discuss designs, with safety and quality being of paramount importance. As the effect of the alcohol waned they remembered that White Star could only finance two ships at the time, so the third would have to wait until later. Back at the Belfast shipyard, Pirrie, who was to hold his post as chairman until 1924, debated the idea with his team of designers: Thomas Andrews, managing director and head of the design department, his deputy (and Pirrie's nephew) Edward Wilding and shipyard managing director Alexander M. Carlisle.

A fixed contract of £3 million for the construction of *Olympic* and *Titanic* was signed and the two ships were allocated yard numbers 400 and 401 respectively. But although the Harland & Wolff Shipyard was the largest in Europe, such was the size of the project that problems concerning its existing facilities quickly came to light. The company owned separate yards – Queen's, Musgrave, Abercorn and Victoria – and a decision was made to replace three slipways at Queen's with two larger ones. The new slipways would need to be equipped with the biggest and most powerful cranes ever constructed, and a huge gantry that incorporated a set of mobile cranes was erected over building slips 2 and 3. But Pirrie was planning further ahead, purchasing a second-hand German floating crane with a lift of 250 tons and organising the building of a new graving dock that would accommodate the liners after their completion. (The term 'graving' is intended to describe a dry dock whose shape resembles a human grave.) Finally, a piece of the riverbank opposite the slipways was cut out as the ships would be longer than the river's width. This later became known as 'Titanic Cut'.

The two liners would, therefore, be built side-by-side, the construction of the second ship being a duplication of that of the first. They had thirteen months until the launch date, then a further five months for fitting out. *Olympic* was laid down on 16 December 1908, but as many of the shipyard's alterations had yet to be finished, she was not launched until 20 October 1910. Without celebration or even a naming ceremony, that strange policy of the White Star Line, she quickly took to the waters, her hull painted white for photographic purposes. She would be officially named at the Atlantic Dock, Southampton, White Star's dedicated berth, following her completion. *Olympic* was a floating palace with the most luxurious shipboard facilities of the day, exactly how Ismay and Pirrie had envisaged her. She even boasted the first 'floating' swimming bath, had squash courts and eleven different styles of decor adorned her first class staterooms. Measuring 274.3m in length and 28.65m in width, her gross tonnage as built was

45,324. *Olympic* began her maiden voyage to New York on 14 June 1911 under the command of Captain Edward John Smith, carrying among her passengers the design head, Thomas Andrews. Fortunately, the voyage was without incident, but several other mishaps impaired *Olympic*'s early years, in particular a collision with the Royal Navy cruiser HMS *Hawke* outside Southampton in September 1911. The liner returned to Belfast for repairs which necessitated the transfer of a number of shipyard workers from *Titanic*, thus postponing the second sister's sea trials. But for *Olympic*'s accident, would *Titanic*'s fate have been happier, for, by delaying her completion by one month, *Titanic*'s maiden voyage took place not in March 1912, as planned, but in April, a month when North Atlantic icebergs tended to prevail? We shall never know.

The keel of *Titanic* was laid on 31 March 1909, whilst her sister was still under construction. As already intimated, her design virtually mirrored *Lusitania*'s, although there were a number of minor differences between the two vessels upon completion. But in view of the controversy surrounding her sinking that has not only continued but escalated to this day, forming the very core of numerous books and films, we should cast a close eye over her design and construction methods.

By employing the constructional format of keel, frame and shell plating, *Titanic* was assembled in exactly the same way as *Lusitania* and *Mauretania*. Her centre girder keel, which was effectively her spine, was a hollow section, 1.6m deep, mounted on a flat keel plate 4.5m thick. Placed on this keel plate and forming the basic hull strength member was a solid steel bar just under a metre in thickness. In addition, four longitudinal girders extended along the bottom of the hull. They were not continuous from stem to stern due

Four-funnelled grandeur. *Olympic* entered service before her ill-fated sister *Titanic*.

to the width of the machinery space compartments that had their own girders of similar strength. Upright columns reached from the girders to the third deck, above which they were replaced by solid steel pillars. It was clear, therefore, that considerable concentration had been placed on providing hull strength. Like the Cunarders of 1907, *Titanic* was constructed with a double bottom, its tank carrying water ballasts that would flow along the ship.

From the vast double bottom rose side frames, forming the now-familiar skeleton effect, to which shell plating was attached. These frames extended as far as the base of *Titanic*'s Bridge Deck, some 20m high, and were evenly spaced at intervals of just under a metre (3ft), although their frequency was increased slightly at the bow and stern. The shell plating was 2.5cm thick, again similar to that used on the Cunarders. They were of standard dimension, being just over 9m long and around 2m wide, precisely the same dimensions as the plates used for the deck structure. Deck beams, of the same depth as the side frames, were then attached to each frame. With the skeleton continuing to grow, transverse beams were placed at each deck level, being attached to the side frames by bracket plates reinforced with angle bars.

From the structural material so far assembled, it is clear that both White Star Line and Harland & Wolff had paid more than adequate attention to the safety aspect of the ship, particularly in providing sufficient strength to withstand the rigours of typical Atlantic winters. But, as the debate over the real reason for *Titanic*'s fate still rages on, a design problem concerning the next components to be installed, the ship's watertight bulkheads,

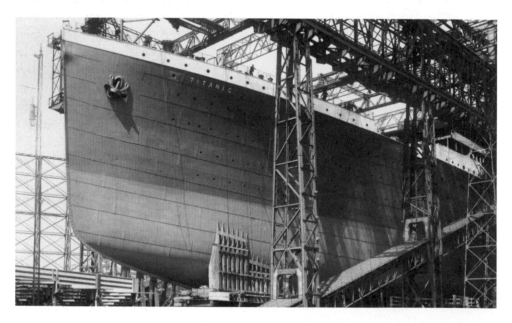

All plates and rivets. The bows of *Titanic* protrude beyond the length of the overhead gantry as the day of her launch draws near.

proved, until recently, the most likely constructional reason for the liner's demise. Fifteen transverse watertight bulkheads, extending from the double bottom to the Upper Deck at her forward end and the Saloon Deck at her aft end, were fitted. As a result, any two compartments, it was claimed, could become flooded without risk to the ship's safety. But they failed to extend as far as the first continuous watertight deck, the Shelter Deck, the uppermost deck to run the full length of the ship, positioned immediately below the superstructure. In accordance with the requirements of the Board of Trade, the bulkheads actually reached further above the waterline than the Board's regulations required, but the damage caused by *Titanic* initially having contact with the iceberg and then scraping herself along it resulted in penetrations that extended over at least five compartments. Consequently, *Titanic* was unable to cope with the extent of water flooding in over the top of her watertight bulkheads.

Thanks to the subsequent location of *Titanic*'s wreck and the technical ability to gain access to it, part of her hull was recovered in 1996 and taken away for tests which showed that the steel used for its construction would have been prone to weakness at low temperatures. In truth, steel of the best quality available was used by her builders and the fact that so much of her hull remained intact even as she reached the seabed was a testimony to its toughness. However, another vital piece of evidence emerged from those tests of 1996. They showed that rivets made of iron rather than steel had been used to attach the hull plating to the framing. At the ship's bow they would have been driven in by hand rather than hydraulically, owing to the curvature of the hull at that point. Further tests proved that the fact that iron rivets used in conjunction with steel plates would have caused increased movement of the plates when under physical pressure. Moreover, the builders had only used the second best standard of iron for the rivets ('best' rather than 'best best'), meaning the rivet heads would have had an element of weakness and very likely 'popped' on contact with the iceberg. So it would appear that, as far as constructional weaknesses were concerned, the reason that *Titanic* sank in just two hours stems from the lack of height of her watertight bulkheads, and also the problem with her hull's rivets.

The story of this great maritime disaster has become increasingly romanticized over the years, but criticism of *Titanic*'s crew has never relented. One of the differences between *Titanic* and her sister *Lusitania* was the enclosure of half of *Titanic*'s forward Promenade 'A' deck with lockable windows, the keys for which had been placed elsewhere and consequently were not to hand as the ship started to go down, delaying access of some passengers to the lifeboats. Other reports have highlighted the absence of any binoculars in the ship's crow's nest and a number of telegraphic errors that prevented ice warnings from other vessels reaching *Titanic*'s master. More than 1,500 lives were lost on that terrible night of 14/15 April 1912, and there was much criticism concerning the inadequate number of lifeboats aboard the liner. However, the finger of blame cannot be pointed at *Titanic*'s owners and builders, as safety regulations implemented in 1894 and still applicable to new ships at the time of her build stipulated that only sixteen lifeboats needed to be carried by any ship of over 10,000 tons. *Titanic* was equipped with twenty, plus five additional collapsible craft, allowing a total capacity of 1,176 persons. Considering that the liner had been designed to accommodate 3,547 passengers and crew, *Titanic* was clearly under-equipped, but by a stroke of good fortune she was nowhere near full, a total of 2,208 persons having boarded for her maiden voyage, otherwise the loss of life would have been truly unthinkable.

Titanic's fate has ensured her prominent place in history forever and tends to overshadow the fact that she was conceived as one of three sisters of differing fortunes. She was 15cm longer than *Olympic*, and with her minor onboard differences enjoyed the accolade of being the world's largest ship, at 46,329 gross tons, on completion. A centrepiece of her sumptuous first class facilities was a grand staircase located between her first and second funnels and leading down to 'E' deck. Its magnificent gilded balustrades, oak panelling and ornate dome above were superbly re-created in the 1997 blockbuster film *Titanic*.

The White Star Line took account of *Titanic*'s shortfalls and despatched *Olympic* to Belfast, firstly to increase her number of lifeboats and later in 1912 to raise the height of her watertight bulkheads and add a second layer to her hull. These alterations increased her gross tonnage to 46,439, but she had already re-taken the mantle of the world's largest liner following *Titanic*'s loss. The third sister, whose construction had been intentionally delayed, was laid down on 30 November 1912 and launched on 26 February 1914, her name having been changed from *Gigantic* to *Britannic* to avoid any sensitivity following *Titanic*'s accident.

She was the largest of the three, measuring 48,158 gross tons, but before she could be completed, Britain declared war on Germany and she entered service as a hospital ship, HMHS *Britannic*, on 13 November 1915. Following a number of successful voyages she struck a mine in the Aegean Sea on the morning of 21 November 1916. Despite the fact that she had been constructed with a double hull and her watertight bulkheads reached 'B' deck, three decks higher than *Titanic*'s, severe distortion caused her to sink with the loss of thirty lives out of a complement of 1,066. *Olympic* saw the war as a troopship before reverting to the role of transatlantic liner on the resumption of peace. She was broken up in 1937, ending her career as the only ship of that initial trio of liners dreamed up by Ismay and Pirrie ever to reach New York.

With the Germans and the British reigning supreme in terms of speed and size, it was now the turn of France to come aboard with the building of their largest and most powerful liner so far. Laid down in February 1909 as *La Picardie*, the ship would be fitted with quadruple screws and powered by Parson's steam turbines. Generating almost 45,000-shaft-hp, the turbines would give her a maximum speed of 25 knots. Yet, although she would be France's biggest vessel by some margin, her gross tonnage of 23,769 would be only approximately half that of White Star's *Olympic*-class liners.

There was much expectancy as the new liner steadily took shape at the Saint Nazaire yard of Chantiers de l'Atlantique (Shipyards of the Atlantic), and hundreds of spectators crowded to watch as she entered *la Loire* for the first time on 10 September 1910. She was aptly named *France*, a sensible name-change having been decided upon as she lay on the blocks. Whilst migrant movement between Europe and America was very much at a

The magnificent first class dining room, complete with grand staircase, was a fine example of the Louis XVI opulence found on the *France* of 1912.

high, the travelling elite of the day desired to be pampered in wall-to-wall opulence, and it was upon this aspect that the liner's owners, Compagnie Generale Maritime (CGT or, more popularly, French Line) were pinning their hopes of financial success.

The *France* would prove a match for the majority of transatlantic liners with her speed – only those veritable greyhounds *Mauretania* and *Lusitania* could reach New York more quickly – and what she lacked in dimensions she certainly made up for in sheer luxury. She became known as the 'Château of the Atlantic', although perhaps 'Floating Palace' might have been a better description, for her Louis XIV decor was truly something to behold. The saying 'no expense spared' is often used flippantly, but in the *France*'s case no truer term could have been applied. Metre for metre her interior cost could not be matched even by Germany's Louis XVI-style four-stackers or the first class elegance of White Star's *Olympic*. Gilded panelling was in abundance – and to the non-French passenger probably overwhelming. Furthermore, as if to endorse that great French passion, dining, *France*'s designers had featured a three-storey 'Salle à Manger' complete with a grand entrance staircase, a forerunner, as time has shown, to a popular cruise ship concept of almost a century later.

France sailed on her maiden voyage from Le Havre in April 1912, within days of *Titanic*'s loss. Her passengers' feelings as they arrived in New York no doubt bore an element of relief, not only because of the awful news that greeted them, but also on account of their ship's apparent inability to remain stable, even in calm seas. In addition to her annoying rolling, undue vibration was emanating from her propellers, and after a handful of crossings she was temporarily withdrawn from service for the lengthening of her bilge keels and the replacement of her two propellers. Thereafter she was the epitome of dependability, operating alongside *Britannic* as a wartime hospital ship in the Dardanelles and returning to commercial service on the Atlantic and then in cruising. The Great Depression of the 1930s eventually led to her downfall and she was broken up at Dunkirk in 1935.

Evidence of the rising Mewès–Davis partnership featured strongly on Cunard's next passenger ship, the 45,647-gross-ton *Aquitania*, although her interiors were actually accredited to Davis alone. *Aquitania* was a ship of 'almosts'; almost the world's longest liner, almost the world's biggest liner and almost the longest-serving Cunarder in history, beaten only by the venerable *QE2*, as described later in this book. Yet, thanks to her fine reputation and longevity, this magnificent four-stacker deserves a prime place in any ocean liner publication. As Charles Mewès weaved his golden web on a mighty trio of Hamburg America new-builds in Germany, a contentious story surrounded the subject of *Aquitania*'s length, as we shall soon see.

Built by John Brown & Co., Clydebank (for whom she *was* the biggest so far), *Aquitania* was laid down on the same berth on which *Lusitania* had been conceived and where construction of the *Queen Mary*, *Queen Elizabeth* and the *QE2* would later materialise. Launched by Alice Stanley, the Countess of Derby, on 21 April 1913, *Aquitania* was fitted with steam turbines, quadruple screws and, in the wake of the *Titanic* disaster, as

The ever-popular *Aquitania* gave her country thirty-six valuable years of commercial and wartime service.

many lifeboats as it would take to accommodate every passenger and crew member. So many were there that sea views from the Boat Deck were at a premium. But at least her passengers could voyage with the knowledge that their ship was among the safest afloat, and Cunard promoted his new fleet member by advertising just that fact.

Davis had been given freedom within *Aquitania*'s vast deck space to translate his experience of onshore hotel and restaurant designing onto the seas, but he was still only thirty-four and the profound influence that Mewès, very much his older partner, had over his work was no better displayed than in *Aquitania*'s Louis XVI dining saloon. By all accounts, whilst employed on their respective British and German projects, Davis and Mewès were prohibited even from comparing notes, surely an agreement that was extremely difficult to maintain. It was, however, in the design of *Aquitania*'s lounges that Davis really came to the fore, creating some of the most memorable public rooms afloat. The Carolean Smoking Room, panelled with oak and mahogany, an echo of a Jacobean section of the Royal Naval Hospital at Greenwich, a pair of Garden Lounges, much expanded versions of *Mauretania*'s Verandah Café, located amidships, and, surely the grand highlight, the magnificently perfect Georgian Lounge, more famously known over the years as the Palladian Lounge, recalling the delightfully harmonious work of sixteenth-century Venetian architect Andrea Palladio, subsequently embraced in Britain by Inigo Jones and Christopher Wren, were all exemplary.

Aquitania sailed from Liverpool on the first of many Atlantic crossings on 30 May 1914. With war looming she soon found herself serving her country as an armed merchant

The superb decor of *Aquitania*'s Palladian Lounge was inspired by architect Andrea Palladio.

Depicted in her early four-funnelled years, *Arundel Castle* had a career spanning thirty-seven years.

cruiser, hospital ship and troopship, and such was her reliability and honesty that she survived not one but two world conflicts, carrying almost 400,000 persons between 1939 and 1945. From luxury liner to cruise ship to wartime hero, *Aquitania* did Britain and Cunard proud until her final passage to Scottish breakers in 1950.

The story concerning *Aquitania's* length stemmed from Hamburg America's quest to satisfy the Kaiser whose demands that Germany would once again own the largest and longest ship in the world had reached a point of desperation. *Aquitania* was 275.2m long, comfortably exceeding *Olympic's* length and seriously threatening the Germans' intended superiority.

It was 1913 and SS *Imperator*, the first of Hamburg America's latest trio of three-funnelled liners, was ready to leave the Vulcan Shipyard in Hamburg and embark on her trials. An unusual delivery suddenly arrived in the shape of a huge bronze eagle over which was draped a banner emblazoned with HAPAG's motto '*Mein feld is die weet*' (My field is the world). Without delay it was attached to the prow of the new liner, thus adding a further 3m to *Imperator's* length and placating the Kaiser. The whole affair seemed ludicrous, particularly as the second and third members of the threesome would each be built to increased dimensions. But by now there was growing unease across

Where *Titanic* and her sisters were built: Harland & Wolff's Belfast yard no longer constructs ships but is dominated to this day by two massive cranes, *Goliath* (erected 1969) and *Samson* (1976).

Europe, and time would prove that Germany's position at the top of the ocean liner league would be short-lived.

During the early 1920s the Union-Castle Line would be introducing a pair of four-funnelled sisters on their Southampton–South Africa route, the only four-stackers not built for transatlantic service. *Arundel Castle* and *Windsor Castle* each measured around 19,000 gross tons and both had two of their funnels removed whilst refitting during the next decade. *Arundel Castle* gave excellent service until 1958, but her sister was lost in the Second World War. But for her alteration in the 1930s, the former would have been the last four-stacker in liner service, but this accolade was fittingly left for Cunard's veteran of two world wars, the *Aquitania*.

CHAPTER THREE

FIVE DIVERSE DECADES

The outcome of the First World War proved distinctly fortunate for British ship-owners Cunard and White Star as they inherited two of the world's largest passenger ships, eventually allowing the latter company to regain the 'crown' of owning the biggest liner afloat from arch rivals, the Germans. However, this was not before the Kaiser, Ballin and all concerned had enjoyed another spell, albeit very brief, of maritime dominance.

SS *Imperator* was launched at A.G.Vulcan's Hamburg shipyard on 23 May 1912 and boasted the impressive statistics of 52,177 gross tons and a length of 276m as she set sail on her maiden transatlantic voyage from Cuxhaven on 20 July the following year. Charles Mewès had clearly been handed the freedom of Hamburg to design the most opulent interiors imaginable, featuring much eighteenth-century French decor that drew approval even from the Kaiser himself.

On Hamburg America's subsequent two vessels, boiler uptakes to the funnels would be divided in order to allow the centralisation of the main public rooms. Yet, on *Imperator*, Mewès had contrived to create a huge lounge, some 30m long and transversely linking both port and starboard promenades, which were used for dancing and other social activities. Completely devoid of any pillars or columns and looked down upon by a huge glass skylight, almost as expansive as the ceiling itself, the 'social hall', as the room became known, rose to a height of around 10m. Not content with this, however, Mewès introduced a second room of almost equal size that represented a combination of palm court (or winter garden) and dining room: the brilliant French architect had made space for his '*specialité sur les mers*', the Ritz-Carlton Restaurant. Yet, if these two palatial rooms bore the hallmark of Mewès at his best, their magnificence was surely matched by the first class swimming pool, the Pompaian Pool, two decks high and flanked by a spectators gallery, totally influenced by Mewès' Royal Automobile Club in London.

For all her interior grandeur, though, *Imperator* suffered from an acute design flaw that affected her stability in all weathers. At first its severity remained unreported as her passengers basked in Mewès' fine creations, but following her first season her owners were made aware of the profound problem that was much worse than normally unpleasant rolling and could be best described as listing whenever the ship changed her course. She was sent back to the Vulcan Shipyard where it was found that her centre of gravity was too high. To rectify this almost 3m were taken off the top of each of her funnels, her grill room was replaced by an open-air Verandah Café, much of her heavy furniture was replaced by lightweight wicker and 2,000 tons of cement was added to existing ballast.

Hamburg America's next ship, *Vaterland*, was delivered by Blohm & Voss, also of Hamburg, who had taken heed of *Imperator*'s design problems and produced a vessel of more agreeable stability. Surpassing her 'half-sister' in size – she was approximately 13m longer and had a gross tonnage of 54,282 – her handful of Atlantic crossings prior to the outbreak of war validating her claim to being the world's biggest liner. Mewès was again allowed to express himself in the design of *Vaterland*'s huge social hall, Ritz-Carlton Restaurant and Pompaian Pool, a job made easier by the division of the ship's boiler uptakes. By the time the *Bismarck*, HAPAG's final vessel of the trio, had descended the slipway at the Blohm & Voss Yard on 20 June 1914, Germany was on the verge of war and the liner, the longest of the three, was to lay unfinished. Mewès had drawn up his plans for the liner's interior but sadly they proved to be his swansong, for by the end of the year

Formerly Hamburg America's *Vaterland*, the American liner *Leviathan*, 59,956 gross tons, sailed the Atlantic until 1937.

In wartime gloom, the USS *Imperator*, snatched by the Americans as a war prize, lies at New York awaiting a new commercial future.

the great man had died. His work for the Hamburg America Line had taken him to the pinnacle of his career, yet not once in his life did he travel by sea to America.

During the four years of war, only *Vaterland* saw action, seized by the United States Shipping Board in April 1917 and deployed for troop-carrying duties under the name *USS Leviathan*. Laid up in 1919, she had to wait until 1922 for the good news that she would be refurbished as a United States Atlantic liner. SS *Leviathan* operated in that role with much financial success (despite being an alcohol-free ship due to her American registration) until the Great Depression started to bite and she was sold to British breakers at Rosyth in December 1937.

As *Vaterland* had become a valuable war prize, so *Imperator* and the incomplete *Bismarck* were handed to Britain, but not before a brief wrangle between the two nations over the ownership of the former that had appeared across the Atlantic as USS *Imperator*.

The first of the trio built for Ballin, who died in 1928, was taken over by Cunard and renamed *Berengaria*. As she flew the Red Ensign for the first time she must have seemed more like a dummy prize to Cunard, for her instability problems reappeared to the great alarm of her passengers, and once again alterations had to be made. A restaurant became a ballroom, first class marble bathtubs were replaced with lighter metal baths and a further 1,000 tons of pig iron were added to her already increased ballast. *Bismarck* was destined for the White Star Line who fared better with their acquisition, although Blohm

& Voss, not best pleased at being instructed to complete Germany's latest jewel for British owners, handed her over painted in Hamburg America colours and with her German name emblazoned on bow and stern. She was officially renamed *Majestic* and, alongside *Berengaria*, happily ploughed her Atlantic furrows for the remainder of the '20s, until, like *Leviathan*, both ships lost trade during the Depression years.

On completion, *Majestic* had been fitted out as an oil burner and proved highly efficient, her huge popularity causing her to be dubbed 'Magic Stick'. One by one other liners were converted to oil during the early post-war years, ensuring considerably faster port turn-arounds and condemning the backbreaking job of stoker to history. Indeed, the 1920s witnessed several radical changes to the passenger liner industry, particularly in relation to Atlantic trade. In 1924 new immigration restrictions were imposed by the United States of America, resulting in a significant drop in demand for third, or steerage, class berths. Many ships were suitably altered with the introduction of a new tourist class.

But perhaps smaller ships were now the way forward. Cunard certainly thought so, ordering eleven liners no larger than 20,000 gross tons. But time would prove that the romance of travelling on a large passenger ship would never wane and in fact even bigger ships were just around the corner. In the meantime, though, the 56,551-gross-ton *Majestic* remained the world's biggest liner until 1935.

White Star's prize acquisition *Majestic* (ex-*Bismarck*) makes a fine sight as she undergoes an overhaul at Southampton.

The glamorously appointed *Ile de France* became a two-funnelled liner in her post-war years.

As the 1920s bade farewell to the austerity of steerage class and billowing black smoke discharged by coal-fired liners, they said *bonjour* once more to those wonderfully ornate interiors first announced on the liner *France*. It was in the year of her build, 1912, that the French Line and the French Government had come to an agreement that stipulated the construction of four further transatlantic liners, but they could never have envisaged how long this arrangement would actually take. But, as is evident from the interior design of the *France*, the French hardly did things by halves and the initial two ships of the four comfortably reflected her opulence. The *Paris* was lying half-built on her slipway when war was declared, and although launched at the Chantiers de l'Atlantique Yard in 1916, her completion was delayed until 1921. Measuring 34,569 gross tons, easily France's biggest liner at the time, she could accommodate 1,930 passengers and featured several 'added luxuries', including the provision of a telephone in every first class cabin. Having established a successful transatlantic partnership with *France*, the liner found herself consigned to cruising in the depressing '30s, only to have her career unexpectedly cut short. On 18 April 1939 she caught fire whilst berthed at Le Havre, capsizing and sinking in the harbour where she remained for almost a decade.

Although the French had a long-term plan for forming their own Atlantic liner fleet, they showed no interest in building the vessels in pairs, as sisters. The *Ile de France* of 1927 resembled *Paris* in that she had three funnels (the third was a dummy), but there the likeness ended. Considerably larger, the 43,153-gross-ton liner entered service amid a grand fanfare of French publicity. Decorated throughout in an early Art Deco style and fitted out with an unprecedented number of top-class suites, she was an instant hit with well-heeled Atlantic travellers. She was modern-day France afloat, a vessel that took shipboard glamour to a new level. Furthermore, she proved a long-term asset to the French Line, giving the British Admiralty, to whom she was loaned in 1940, invaluable Second World War service, and receiving a new two-funnelled profile during a post-war refit that would extend her commercial life until 1959.

Before the 'roaring twenties' were out, there appeared over the horizon a pair of liners that, although by no means the biggest of their time, were immensely significant. *Bremen* and *Europa* rose from the ashes of Germany's passenger ship fleet, thanks largely to financial assistance awarded by the United States of America in return for vessels confiscated during the war. Their arrival stunned the rest of the shipping world, causing it to sit up with sudden interest at their comparatively squat funnels, sleek hulls and novel bulbous bows. Built for North German Lloyd by A.G. Weser, Bremen, and Blohm & Voss, Hamburg, respectively, the ships were launched on successive days in May 1928, but delivery of the 49,746-gross-ton *Europa* was unexpectedly extended by a year as a result of a serious fire whilst fitting out that almost destroyed her. Yet it was this sister that would go on to live a full and varied life, handed to France after the Second World War and becoming one of the best loved French liners to have sailed the Atlantic. More of

Built with unusually squat funnels, North German Lloyd's *Europa* forged a second career under the French flag.

this handsome ship later. *Bremen*, slightly larger at 51,656 gross tons, saw her career halted in 1939, similarly ravaged by fire whilst lying at Bremerhaven as a barrack ship. She was dismantled in 1946. So, once again, Germany had lost its most prestigious liners as a result of war. But at least they had achieved their intended goal, for both these speedy ships, designed to operate at 27.5 knots, had managed, in turn, to reclaim the Blue Riband, for so long held by the illustrious *Mauretania*.

In the coming 1930s, ocean liner size was to reach a new pinnacle. As we acknowledged earlier, though, planning a huge passenger ship can take several years, and the work behind the building of the next breed of 'superliner' was certainly no exception. The story of the two mighty *Queens* began around the middle part of the '20s when Cunard decided to build a pair of ships that would be both larger and faster than any other passenger liner, past or present. They would not only act as replacements for the ageing *Mauretania*, *Berengaria* and *Aquitania*, but enable Cunard to open its first weekly transatlantic passenger and mail service. The plans for the first, the *Queen Mary*, began in 1926. It was a completely new experience for her designers, for, with a length exceeding 300m, she would be like no other ship so far constructed. Indeed, she would be so large and so expensive that even the British insurance markets could not cover the vessel's full value. So the British Government, following the passing of a special insurance act, agreed to cover the huge sum involved. Even more than this, larger port facilities and deeper channels needed to be created in order to cope with the ship. New piers were constructed at Southampton, Cherbourg and New York, and a new graving dock named after King George V was built at Southampton, principally with the new vessel in mind.

The two Queens would be built at the Clydebank Shipyard of John Brown Co., and the keel of Yard No.534, as *Queen Mary* would be known right up to the day of her launch, was laid down on 27 December 1930. Over the next twelve months the ship grew into an immense steel skeleton as construction proceeded ahead of schedule. Then the Great Depression inflicted its stunning blow on much of the industrial and commercial world, including Cunard Line, who calculated that insufficient funds would be available to complete their new ship. A fruitless plea was made to the British Government who had in no way subsidised the liner and, with heavy heart, Cunard found it necessary to suspend all constructional work, making some 3,000 shipyard workers unemployed within a matter of hours. Across Britain at least a further 6,000 workers involved in sub-contracting industries ranging from steel and electrical to furniture manufacturing found themselves without work. For twenty-seven months the liner sat in eerie silence, accumulating 130 tons of rust and a multitude of birds' nests. A huge structure that had been seen as a beacon of hope for the British people now represented the doom and gloom of the worst economic period in recent history. Furthermore, Yard No.534 was not the only major maritime factor to fall foul of financial problems. White Star Line had been planning to lay down their own version of a 300m liner in 1928, but the company was already losing money and the onset of the Depression meant that construction of their new *Oceanic* would never go ahead.

BUILDING THE BIGGEST

On the other side of the English Channel, Compagnie Generale Transatlantique (CGT, or French Line) had been concerned for some time that, because their fleet of ships was built for comfort rather than speed, their chances of imposing stiff opposition against their rivals would quickly diminish. The fact that the Germans were rejoining the act was concerning enough, but as the news of Britain's proposed new giants filtered across 'La Manche' they knew that they had to act swiftly and positively. Sadly, Henri Dal Piaz, the French Line's inspiration and vision for so long, died in 1928, but his very capable successors took forward plans to build France's first 300m liner that would be more than capable of outpacing her British competitors. Subsequent events would see the French winning on all fronts. With the withdrawal of White Star's plans, only the future *Queen Mary* remained as immediate competition, and her delayed completion would see the new French liner enter service before her. In fact, although she is still to this day regarded by some shipping aficionados as the most famous British liner in history, not once did *Queen Mary* hold the accolade of sailing as the world's biggest ship.

As if the misfortunes of its British rivals were not enough, CGT received a further piece of luck as, out of the blue, revolutionary ideas for the design of their new ship's hull arrived at the Saint Nazaire shipyard from a former Russian naval engineer who was living in France. Vladimir Yourkevitch had created a novel hull design intended for a class of new battle cruisers whilst under employment at St Petersburg (then known as Leningrad). However, the advent of the Russian revolution led to the cancellation of the warships and caused Yourkevitch to flee his homeland. At first his ideas were ignored, but a little persuasion from an influential colleague enabled him to visit the shipyard and impress his hosts.

In layman's terms, his hull design differed from the more traditional form of, for instance the Queens or the *Mauretania* of 1907, in that its sides remained straight for a longer proportion of its length before slanting in more profoundly at stem and stern, thus allowing a predominately flat bottom that would aid stability. To balance this appearance and to aid speed through the water, a clipper-like bow curved gracefully down to a bulbous forefoot beneath the waterline, similar to that featured on the German liners *Bremen* and *Europa*. The forefoot, or bulbous bow as it is popularly called today, helps to make the ship go faster by forcing a hole in the water ahead of the oncoming hull. Coupled with a trio of funnels, aesthetically designed with a ten-degree rake and consistently diminishing in height from forward to aft, the hull's clean lines ultimately produced one of the most beautiful classic liners ever built.

During early days of construction the vessel was referred to simply as *T-6*, the *T* being short for 'Transat', in turn an abbreviation of Compagnie Generale Transatlantique, and *6* meaning she was their sixth ship. After much discussion her owners announced that '*le paquebot nouveau*' would be named *Normandie*. The liner was officially named on 29 October 1932 in front of 200,000 spectators. As fitting out proceeded it was soon apparent that her elegant external profile would be equally matched by her lavish interiors. Divided boiler uptakes, as on Ballin's *Bremen* and *Europa*, allowed ample scope for some wonderfully spacious public areas, none more so than the first class dining room, at 97m long, 14m wide and approximately 8.5m high, comfortably the largest room afloat. Other notable areas included the Grand Salon that would double as a nightclub, and the indoor and outdoor swimming pools, the latter measuring 30m in length. Yet, amazingly, *Normandie* had been designed to carry no more than 1,972 passengers, just over 300 more than the comparatively tiny *France* of 1912.

Below the waterline, powerful turbo-electric machinery was installed, producing up to 200,000hp and a top speed of over 32 knots. The ship's four three-bladed screws weighed 23 tons each. The French Line was determined that *Normandie* would be the largest and fastest ship ever seen, and was meticulous in ensuring that nothing fell short of the very best quality. Last minute alterations were made and the fitting out process lengthened from two to three years. From top to bottom, the huge liner was a hubbub of perpetual motion whilst, on the banks of the Clyde in Scotland. Yard No.534 sat motionless awaiting her future. Conflicting circumstances, for sure, that provide us with the opportunity to catch our breath for a moment and take an overall look at the art of shipbuilding as carried out during the first half of the twentieth century.

Generally speaking, shipyards evolve at coastal locations or on river estuaries, although a few yards, such as Napiers at Millwall in Brunel's time have appeared much further from the sea, often necessitating sideways launchings. However, the choice of a suitable site can include many other considerations, the first being an adequate supply of steel. It is noticeable that Britain's biggest yards, most of which are now defunct, grew up in northern counties within arm's length of its principal steel works from which plates and sections could be readily obtained. An exception was Harland & Wolff's Belfast yard, to which steel had to be shipped across the Irish Sea from Scotland as Northern Ireland was without any local steel works.

Before the days of welding and oil-fired furnaces, coal and coke was required by the proverbial bucket-load for numerous shipyard processes ranging from the work done in huge frame furnaces to that carried out in small rivet fires. As these old fashioned procedures were gradually replaced by electric arc welding and power driven machines and tools, the necessity for readily available electricity supplies became increasingly vital. Additionally, supplies of gas, petrol, oil, fresh water and materials such as timber needed to be to hand.

No shipyard could, of course, operate without a plentiful supply of labour, both skilled and unskilled. The importance of having the correct type of labour within the shipbuilding industry cannot, even in today's world, be over-emphasised, and in the years after the Second World War when many new techniques came into being, the industry required skilled and alert tradesmen as never before. But they needed the provision of the best heavy equipment: large tower or travelling cranes positioned alongside slipways and fitting out berths. As we saw in Chapter Two, Harland & Wolff were required to specially erect giant gantries to facilitate the building of the three *Olympic*-class ships in the early 1900s, and similar equipment is employed nowadays for the positioning of huge prefabricated sections within a ship's building dock.

Shipyards were effectively giant assembly points where the output of many branches of engineering, such as steel mills, foundries and machine shops, were concentrated. Other tradesmen – shipwrights, plumbers, fitters, joiners, electricians and painters, for instance – would contribute their share to the construction of a ship. The focal point of all this great activity was the building slip where the ship took shape and from which the launching of the enormous mass of steel took place. For this reason, the slip needed to be constructed with great care and the ground suitably strengthened to carry the large concentrations of accumulated weight as the building proceeded. This strengthening was achieved by careful piling with cross bracing to ensure the whole area could support the ship's weight. Providing the piling was adequate, the slip was well designed and its construction well supervised, the building berth could last indefinitely.

All slips were built with a certain declivity to ensure the successful launching of the ship. In fact, once released, the ship really launched itself by sliding down the prepared inclined plane, and it was consequently vital that any friction would be successfully avoided. Use of suitable lubricants between the standing and sliding launchways would ensure that once the ship began to slide she would not come to rest until fully waterborne. So when at all possible, large passenger liners would be launched in the warmer months of the year to ensure sufficient lubrication.

After the launch of a new vessel, there were countless jobs for the shipyard workers to do in clearing away all the debris from the slip. This would entail the moving of piles of drag chains, standing ways, large amounts of timber and various other materials. Then followed the building up and lining up of keel blocks, massive pieces of timber, ready to carry the keel of the next ship to be built on the slipway. The top level of these blocks, when finally lined up, had to have a predetermined declivity, requiring painstaking and accurate setting up and alignment. Following this important task, the actual laying of the next keel could take place.

Within the previous two chapters we observed the formation of giant skeletons that would represent the framing of a ship's hull. However, in the construction of Brunel's first iron ships, and later the *Titanic*, special emphasis was placed by the ships' respective designers and builders on keel and hull strengthening. Indeed, it was a popular perception outside the shipbuilding world that the keel was really a solid steel bar reaching from bow to stern. But, in truth, the keel of a typical slipway-built merchant vessel consisted of no more than a series of flat steel plates up to 2.5cm or more in thickness, suitably riveted, or, in later years, welded together and shaped into a gentle curve at the forward and aft ends of the long centre portion. The dishing of the plates, as it was called, was carried out so that a recess was formed to accommodate the stem bar and stern framing. These huge steel castings and forgings would then be joined in place later in the ship's construction.

Steel plates also formed the vertical keel, attached to the flat keel plates already laid on the keel blocks. To this vertical keel transverse floor plates would be attached at regular intervals, extending on each side of the keel to form the vast honeycomb effect of the ship's double bottom. This part of the vessel would later contain the main water tanks, constituting the whole of the double bottom, mainly for carrying water ballast. When lightly loaded, with little fuel, cargo and fewer passengers, a ship floats higher and can be potentially unstable. So the tanks are filled with water to overcome this. When rough weather is expected a ship takes on extra water ballast to counteract the effect of heavy seas and strong winds.

The process of hull framing involved the heating of long steel bars of channel or angle sections in furnaces within the platers' shed that were then bent to various predetermined shapes for the different sections of the ship. Frame setting would have been a fascinating operation to watch and would be followed by the riveting of brackets, to which the deck beams would later be connected, to the frames. Once the skeleton had taken shape, transverse beams were placed athwartships and connected to the beam knees on their respective frames. The beams were also supported fore and aft by metal girders and vertical pillars, and as the work proceeded the various decks were laid. The steel, or shell, plating was then attached to the frames to form the huge watertight structure that was the ship's hull. It is worth noting that the same standard size of shell plating – 9.1m by just less than 2m – used on ships in the early 1900s was still being supplied by manufacturers some fifty years later, although the plates would have been of somewhat improved composition.

Today's method of shipbuilding entailing the use of prefabricated blocks has taken away much of the hard slog suffered by those hardened workers who created ships in the days when Britain dominated the worldwide shipbuilding industry. Yet there is little doubt that much of the romance has disappeared from the procedure with the elimination of the dramatic slipway launch, while keel construction has been replaced by the simple laying of a prefabricated section of a ship's double bottom (and not necessarily the most central portion) on the floor of the building dock.

Ready at last! *Queen Mary*, sporting a white hull for photographic purposes, sits proudly on her Clydebank slipway prior to launch.

As we return to the differing fortunes of the future *Queen Mary* and France's new flagship *Normandie*, good news at last prevailed for the new Cunarder. Following lengthy negotiations, the British Government finally agreed to advance the sum of £4.5 million to Cunard and promised a further £5 million when the second of the Queens was built. But there was one important proviso: that Cunard would take over the assets of the ailing White Star Line to form a new single company. This was a perfectly sensible move, for it not only precluded the necessity for White Star to apply for Government help itself, but awarded Cunard with £3 million to finish Yard No.534 and £1.5 million as working capital for Cunard White Star Ltd, the newly formed company, as well as a guarantee that a further huge liner would be built. On 3 April 1934, 400 men marched through the gates of the John Brown shipyard to the tune of 'The Campbells are Coming', heralding the resumption of work.

Yard No.534 was launched by HM Queen Mary on 26 September 1934, the first time a reigning British queen had named a merchant ship. Accompanied by King George V and the Prince of Wales, she watched the 35,500-ton hull slide down the slipway on 200 tons of tallow and soft soap. The River Clyde had been specially widened to accept the new liner. King George V spoke at the ceremony, stating: 'It has been the nation's will that she should be completed. And today we can send her forth no longer a number on the books but a ship with a name in the world, alive with beauty, energy and strength.'

Back at Saint Nazaire, *Normandie* was nearing completion and, following successful trials, was proudly accepted by the French Line as their new superliner. Fifty thousand people gathered to watch her depart on her maiden voyage from Le Havre to New York on 29 May 1935, hoping that they were witnessing the beginning of an historic

Queen Mary takes to the waters of the Clyde for the first time.

Entrée magnifique. Onlookers gather to witness France's biggest ever liner, *Normandie*, as she sails into her home port of Le Havre.

record-breaking voyage. Their hopes were realised as *Normandie* completed her voyage in four days three hours and fourteen minutes. The Blue Riband was at last in French hands, *Normandie* having snatched the prestigious award, not from the Germans, but the Italians, who had joined the transatlantic scene in 1931 with their first liners of real significance.

Conte di Savoia and *Rex* entered service two months apart in 1932, having been completed at Genoa under the close scrutiny of the fascist dictator Benito Mussolini who had ordered the formation of a new shipping company, Italia Line, from the merger of three separate companies, to operate the ships. The original owners of the larger *Rex* had intended to name her *Guglielmo Marconi*, but the appearance of King Emanuel III at her launch prompted a late change of mind. Both the vessels will forever be remembered for their outdoor lido areas, complete with expansive swimming pool, sand and colourful umbrellas. The *Rex* actually broke down on her maiden voyage, but in August 1933 the 51,062-gross-ton liner took the Blue Riband from *Bremen* with an average speed of 28.92 knots.

Normandie took the award from *Rex* with westbound and eastbound crossings at 30 knots, but there was no time for complacency as Britain's own new superliner was almost ready to take up the challenge. Fitting out on *Queen Mary* took two years, notably quicker than on *Normandie*, her four sets of quadruple-expansion, reduced-gear turbines and twenty-four boilers being installed whilst the finishing touches to her high-class accommodation were being made by thirty artists, sculptors, painters and interior decorators.

At full steam. The liner *Rex* took the Blue Riband for Italy in 1933.

She was certainly a ship of superlatives: 10 million rivets were driven into her hull and superstructure; steel plates, each up to 9m long and weighing as much as 3 tons formed the outer hull and decks; 2,000 portholes and windows were installed; 70,000 gallons of paint were applied to her outer hull; and for the safety of her passengers, twenty-four lifeboats were carried, each capable of holding 145 persons.

One fascinating story that perfectly demonstrates the amount of importance that shipping nations and their respective shipping companies attached to having the honour of owning the world's biggest passenger liner involved both the *Normandie* and *Queen Mary*. On completion, the Cunarder measured 81,237 gross tons compared with the *Normandie*'s 79,280. On hearing this fact, the French Line lost no time in adding an enclosed tourist class lounge to their flagship's aft Boat Deck, and with this cleverly contrived addition and a few other small alterations her gross tonnage immediately jumped to 83,423. Consequently, *Normandie* comfortably retained her position as the world's biggest ship. *Queen Mary* would have to be content with the title of being Britain's biggest liner and the largest ship to have come from a British shipyard – for the time being, at any rate.

In March 1936 *Queen Mary* sailed down the River Clyde to commence her trials, and on 27 May she steamed out of Southampton on her maiden voyage to New York where she received a rapturous welcome. Her predominately Art Deco-style interiors appeared restrained compared with her more sumptuously appointed French rival, but within three months of her first Atlantic crossing she finally 'got one over' on *Normandie* by grabbing the Blue Riband in both directions, achieving an average speed of 30.14 knots westbound and 30.63 knots eastbound. This prompted even greater competition between the liners, *Normandie* reclaiming the trophy in 1937, only for the *Mary* to respond in the following year with an eastbound voyage at 31.69 knots, a record that was to stand for fourteen years.

Queen Mary had entered service some ten years after her first appearance on the drawing board, and by this time plans for Cunard White Star's second superliner were well under way, it being essentially a copy of the *Mary*, but with a number of improvements. The new ship's architects decided that they would learn a lot from the design of *Normandie* and they despatched a representative to carry out an undercover inspection. The man posed as an English grocer going on holiday to America and undertook a thorough examination of the vessel and questioned crew members at every opportunity. There is no doubt that his reports successfully influenced the development of the new liner.

The shape and form of the ship's hull was initially modelled in wax and was involved in almost 8,000 tests that helped to determine its ideal design, including the propellers and rudder. The John Brown Co. tested a 5m model in a long tank, including the simulation of storms created by waves generated within the tank. As a result, the ship's seaworthiness was proved at an early stage, whilst her superstructure and funnels were designed with the assistance of a wind tunnel.

A towering presence. The hull and superstructure of *Queen Mary* was constructed using 10 million rivets.

At a glance, the overall external profile produced was not unlike that of *Queen Mary's*, the main difference being that the new liner would be endowed with two funnels as compared with *Mary's* three. Another notable variation was the elimination of the well deck, in layman's terms the lowering of the main deck immediately behind the fo'c's'le. The *Mary* had been designed as such because it had been considered that the inclusion of this feature would dispel the shock that would fall upon the wide expansive forward part of her superstructure should the ship dip her nose in heavy seas. This fortified well would, it was thought, cause sea water coming on board to lose its power and destructive effect before reaching the superstructure. However, it had been found that *Queen Mary* had pitched no more than a few degrees even during her worst Atlantic crossings and had not shipped any water of note, so the idea was abandoned for the new liner. This resulted in an unbroken clean sweep of the forward deck, which was easier on the eye and permitted increased internal capacity.

The new vessel would look longer and sleeker than the *Mary* through the loss of not only the well deck but also the many ventilator cowls and the network of supporting guy wires for her funnels, all of which gave a cluttered appearance to her upper external profile.

The reduction, by one, of the number of funnels was made possible by improved technology, for, whilst the *Mary* required twenty-four boilers for the operation of her four sets of geared turbines, her new sister would need only twelve. There would as a result be more space for passengers – 2,260 compared with *Mary's* 2,038.

Cunard White Star announced the name of their second *Queen* as soon as building began, and *Queen Elizabeth's* keel was laid in December 1936. Construction plans were laid out on a weekly schedule, special 'shock brigades' of shipbuilding specialists being set up in order to assist those workers responsible for the various parts of the ship to stay on schedule. If one section began to lag behind the others, a 'shock brigade' was rushed to that section to help its workers catch up. In all, 3,000 workers were involved in the ship's construction, and it was estimated that as many as a quarter of a million others took part in the supply of materials from across much of Britain.

On 27 September 1938 *Queen Elizabeth*, now a 40,000-ton hull, was launched, only three days before the completion of the Munich Pact. Because of the situation the King did not attend, although the rest of the Royal Family was there. HM Queen Elizabeth (later the Queen Mother) was to launch the ship, and she was accompanied by Princess Elizabeth (later Queen Elizabeth II) and Princess Margaret. Whilst waiting for the tide to peak, Lord Aberconway, chairman of the John Brown Co., was indicating to the Queen which button she should press to send the giant ship down the slipway when he accidentally touched the button himself, sending the ship on her way. Startled officials, initially speechless, suddenly realised that the liner would reach the waters without being named. By tradition this would leave her nameless, so Lord Aberconway called to the Queen to immediately perform the naming. But the Queen, looking surprised herself, replied that she had already named her, thinking nothing untoward had gone wrong and that she had proceeded to do her part, including the release of a bottle of Empire wine

Undercover escape. All in grey, the *Queen Elizabeth* slips quietly down the River Clyde en route to her secret destination, New York.

which had swung out on a ribbon before smashing against the liner's fast receding bow. Amid cheers from the onlookers the Queen announced: 'I hope that good fortune may attend this great ship and all who sail on her. I am very happy to launch her and name her *Queen Elizabeth.*' But her words went unheard as her microphone had failed to work.

As the new liner slipped gracefully into the Clyde she dragged thousands of tons of chains behind her, enabling her to safely come to a stop in the river. A few of these chains had served at that protracted launch of Brunel's *Great Eastern. Queen Elizabeth* remained in her fitting out basin for almost two years, the date for her maiden voyage having been set as 24 April 1940. But soon the world was at war again and for months she lay unfinished, all work having come to a halt. As Nazi bombs began to rain down on Britain, it became obvious that the nation's biggest liner was an attractive target and that she should be moved without delay.

The 26 February 1940 was a typically dull Scottish winter's day, but it was also a day when the tide was especially high. Slowly but surely the *Queen Elizabeth* was eased into the Clyde where she would pause at anchor. Rumours were rife – deliberately circulated by those in authority – that she would set sail for Southampton, her home commercial port. On the morning of 2 March sealed orders arrived on her bridge, accompanied by strict instructions that they should not be opened until the liner was at sea. Within hours the *Queen Elizabeth*, her all-grey silhouette merging with the skies, sailed out of the

With the world now at peace, RMS *Queen Elizabeth* takes on passengers at Southampton's Ocean Terminal.

Clyde and into the awaiting sea. Her master, Captain Townley, duly opened the orders to find that he was to take his ship to New York. To the south of the liner, Luftwaffe bombers were already flying over the Solent and Southampton where she should have been sailing.

A rapturous welcome greeted the *Queen Elizabeth* on her arrival at New York where she berthed alongside *Queen Mary* and *Normandie*. Both Queens would subsequently be sailing for Sydney, Australia, for refitting as worldwide troop transports, a role in which they proved incredibly successful thanks to their size and speed. Once peace resumed they would go on to become the most famous maritime double act of all time, carrying the rich and famous among their many thousands of passengers until the jet airliner forced them to exit the great Atlantic stage. As for *Normandie*, she was never to leave that New York berth at Pier 88, for during conversion into an American troopship, the USS *Lafayette*, she caught fire on 9 February 1942 and capsized on the following day under the weight of water pumped into her. The salvage operation took fifteen months and the once magnificent liner, the pride of all France, was sold for scrap.

By the end of 1950 the French Line had a replacement, albeit a much smaller one. North German Lloyd's *Europa*, having outlived her sister *Bremen* and survived the Second

World War, was taken to New York in 1945 as a United States war prize for conversion into a troop transporter. But before work began, vital information emerged from German sources that structural cracks had appeared on two occasions after her completion. It transpired that the ship had originally been intended to be shorter than her 285.3m length, and such was the haste to finish her, necessary design changes that would have accounted for her extra length failed to be implemented.

Although conversion went ahead and *Europa* made two voyages under the US flag, the structural deficiencies and additional problems with the ship's electrical system, which was found to be of especially poor standard, made her unsuitable for her new role, and she found her way to France as compensation for the loss of *Normandie*. Several names for the French Line's new acquisition were considered – *Lorraine*, *Liberation* and even *France* – before she was christened *Liberté*, an appropriate name in her circumstances, on 27 July 1946. She was laid up at Le Havre, awaiting reconstruction into a passenger liner, but on 9 December the worst storm in the port's history caused her to break from her moorings and she was blown onto the wreck of *Paris*, which had lain in the harbour since 1939, and she too sank to the bottom.

Liberté was re-floated and towed to Saint Nazaire for conversion into a 51,840-gross-ton transatlantic liner that for a while was the third largest in the world after the two Queens. From her maiden voyage on 17 August 1950 until November 1961 when she was replaced by French Line's new *France*, she was simply *magnifique*, an excellent

A notable feature of the *Queen Elizabeth*'s main public rooms was their high ceilings.

France's '50s flagship *Liberté* (ex-*Europa*) pictured before the addition of enhancing funnel tops.

replacement for the great *Normandie*, her elegant profile being enhanced further in 1954 by the addition of new funnel tops. Renowned for her superb food, she became known as one of the world's top restaurants on the Atlantic, which she traversed 199 times during her French career.

With the 1940s at an end, the 83,637–gross-ton *Queen Elizabeth* was well established as the world's largest liner, with the *Queen Mary* not far behind. Yet, although their respective schedules demanded similar speeds, the *Queen Elizabeth* never did hold the Blue Riband. With the sad demise of *Normandie*, though, Britain would surely keep the trophy safe in its hands... or so it thought. In 1952 along came America's challenge in the streamlined shape of SS *United States*. At 53,329 gross tons, considerably smaller than the Queens but clearly America's biggest liner at the time, the ship was the inspiration of William Francis Gibbs, a highly gifted maritime architect who had designed the vessel's older and smaller running-mate, SS *America*, for the United States Lines. He had impudently spied on *Normandie* whilst she was docked at New York, but took his designs to completely new heights. The *United States* was constructed by the Newport News Shipbuilding & Dry Dock Co. in a building dock, resting on horizontal keel blocks and incorporating numerous prefabricated parts that could be easily on-loaded, as workers enjoyed easier access to the ship than if she had been built on a conventional slipway. By the time she was christened and floated out on 23 June 1951, the liner already possessed her engines and boilers, her bridge and even her two mighty, distinctive funnels.

Gibbs was obsessed with the matter of fire safety and had installed considerable amounts of aluminium within the ship, including her superstructure, thus reducing her overall weight. By coupling this concept with massive, powerful engines, he had created a veritable Atlantic greyhound. On her maiden voyage from New York, the *United States* steamed from the Ambrose Lightship to Bishop's Rock in three days, ten hours and forty minutes at an amazing average speed of 35.59 knots. Her westerly crossing averaged 34.51 knots, making her the new undisputed holder of the Blue Riband and the last liner ever to hold the award.

As the 1950s progressed, the eyes of the shipping world turned away from the competitive Atlantic to the comparatively sedate southern hemisphere. The Union-Castle Line had run a regular passenger-mail-cargo service between the United Kingdom and South Africa for half a century. Their vessels' weekly departures from Southampton were renowned for their punctuality, so much so that it was said that local residents would set their watches by the ships' sailings. In 1956 details of the company's biggest ship, and also, as events transpired, their last, were released. She would be built at the Birkenhead yard of Cammell Laird, breaking a long-term tradition of employing Harland & Wolff to construct their vessels. As the handsome ship took shape it was announced that she would be named *Windsor Castle*, a successor of the 1922-built ship of that name, and she was launched, appropriately, by HM Queen Elizabeth The Queen Mother on 23 June 1959.

Designed to carry 250 first class and 600 tourist class passengers, the 37,640-gross-ton liner had space for some 70,000cu.m of cargo. Entering service on 18 August 1960, she

America's sleek Atlantic greyhound, SS *United States*, departs from Southampton.

gave seventeen years of reliable service to Union-Castle, completing 124 round voyages that would feature calls at Las Palmas, Cape Town, Port Elizabeth, East London and Durban, carrying a total of 270,000 passengers. Unfortunately, faced by stiff competition from long-haul flights and the containerisation of cargoes, *Windsor Castle* was deemed uneconomical by 1977 and was sold to Greek owners, later to be laid up at Eleusis until sailing for demolition on the beaches of Alang in 2005.

P&O is one of the longest-existing shipping concerns, and for a brief while in the mid-1800s owned the world's biggest passenger ship, the 3,438-gross-ton *Himalaya*. In later years, in partnership with Orient Line, the company fairly dominated the routes to Australia, particularly at times of high migration before and after the Second World War, yet neither company owned a passenger vessel exceeding 30,000 gross tons. By the 1950s both concerns were extending their services across the Pacific and recognised that they needed larger and faster liners that would reduce the passage to Australia from four to three weeks, thus making the whole round voyage a more agreeable length of time. Orient Line were first to order their ship, and on 18 September 1957 the keel of Yard No.1061 was laid at the Vickers-Armstrong Shipyard at Barrow-in-Furness. *Oriana* was launched on 3 November 1959 by HRH Princess Alexandra and entered service from Southampton on 3 December 1960. On her eleven decks the liner could accommodate 2,184 passengers, just thirteen fewer than RMS *Queen Elizabeth*, which was over 60m longer. On her trials, the 41,915-gross-ton *Oriana* achieved 30.6 knots, and with her 27.5-knot service speed she could easily have held her own on the Atlantic run.

Determined passengers brave the North Atlantic chill as the SS *United States* makes another speedy crossing.

A legend in her time. The P&O liner *Canberra* carried troops to the Falklands before becoming a successful cruise ship.

The new superliner planned by P&O was to exceed her compatriot in tonnage and length, but not in speed. Ordered from Harland & Wolff in January 1957, the ship would be the very last liner built at the Belfast yard and was launched as *Canberra* by Dame Pattie Menzies, wife of the Australian Prime Minister, on 16 March 1960. The sleek white ship with her novel engines-aft design entered service on 2 June 1961, departing from Southampton with a full complement of 2,238 passengers, including 750 migrants bound for new lives in the Antipodes. Both *Canberra* and *Oriana* fully justified the massive investment that their respective owners had put into their construction, and from 1966 were sailing under the same P&O banner after the Orient Line had become wholly owned by its fellow operator. *Canberra* went on to attain legendary status through her participation as a troop transport in the 1982 Falklands War and later as a highly popular cruise ship. *Oriana* found her cruising career curtailed and she served as a static floating attraction in China from 1987 until she was sent for breaking in 2005. Both the vessels possessed their own unique external designs that will be more closely explored in the final chapter.

From the 1960s relentless competition from the jet airliner offered no sympathy to the ocean liner, and Cunard shook the maritime world by announcing the withdrawal of *Queen Mary* and *Queen Elizabeth*. In 1967 *Mary* steamed for Long Beach, California, where she has remained ever since as a hotel and museum. *Queen Elizabeth* ceased transatlantic service a year later and by 1970 was in the hands of Hong Kong tycoon C.Y. Tung for conversion into a floating university. But tragedy struck in January 1972 when, during the refit, the world's biggest passenger ship was ravaged by fire, believed to be an act of arson, and toppled into Hong Kong's Victoria Harbour. In little time at all two 80,000-ton liners had left our oceans, and not until a new era of cruising was in full swing would passenger ships of comparable size be built again.

CHAPTER FOUR

RE-BUILDING THE BIGGEST

'I'm so sorry about the news,' sympathised a concerned-looking American passenger as I was leaning on the ship's rail.

'Yes, I'm really sorry,' called a crew member.

'That's all right,' I muttered, looking back at both of them incredulously.

I had risen early, a tingle of excitement in my veins. My plan was to be on deck, camera in hand, as we sailed into Le Havre. A quick glance through my cabin porthole had revealed that a flotilla of small craft had already caught up with our ship, so I hastened up several staircases to the Sun Deck.

I sought out a friendly face, a comfortable-looking middle-aged lady.

'What's happened?' I asked.

'It's Princess Diana. She's died in an accident,' she explained, 'in Paris.'

It was Sunday 31 August 1997. The cruise ship *Norway* was amid a rare return visit to European shores and many fellow passengers were booked on a shore excursion to Paris that day. I, however, had opted to stay closer to the ship despite the fact that she would be remaining in port until later the following day.

I shall linger no longer on the events in Paris that previous night, as so much has been said since. Suffice to say that conversation on board during the next couple of days continually re-directed itself away from the topic of sea travel.

With helicopters overhead and busy little boats daring to encroach too closely to *Norway*'s massive dark blue hull, we swung around within Le Havre's expansive harbour and berthed ahead of Holland America Line's *Maasdam*, a cruise ship of modern profile that contrasted perfectly with the traditional sweeping lines of our much older vessel.

The SS *Norway* dated back to 1961. As many passenger ship experts will know, she was built as SS *France*, the largest French liner since the *Normandie*, and she plied the North Atlantic at a time of increasing competition from the skies above. But what shipping fans

Bienvenue à France! SS *Norway* (ex-*France*) arrives off Le Havre in 1996, her first visit to her former home port for twelve years.

may not appreciate is the genuine affection the people of the French nation held for her, even though she had spent more years of her career based in America than flying the French flag.

Norway had made a similar sojourn into Europe during September of the previous year. Her first fare-paying transatlantic crossing for twelve years, the voyage had been arranged to coincide with a $4.8 million refit at Southampton to bring the ship into line with impending SOLAS requirements. *Norway* had embarked passengers at New York, and with calls at Southampton and Le Havre, the voyage had been made into a celebratory affair, evoking memories of the days when she would make the same journey as France's premier liner. The idea proved such a success that her owners, Norwegian Cruise Line (NCL), decided to repeat the exercise, but this time the whole round voyage would be marketed as a journey of nostalgia.

I had been invited to join the ship at Southampton for the cross-channel portion of the voyage. Cabin V045 was equipped with a double bed (especially comfortable for a single occupant), private bathroom and was reasonably spacious without being pretentious. The room was marketed as an Oceanview Stateroom, Category D. Above the dressing table was fixed a framed poster commemorating the ship's French Line days, and nearby I was embarrassed to find a personal certificate in my name, certifying my transatlantic crossing on the former *France*, a voyage I had, of course, not made. My thoughts turned to the ship's origins and the reasons why she had endeared herself so much to the people of France.

It was during the 1950s that the CGT (French Line) decided it was high time it ordered new tonnage to replace the liners *Ile de France* and *Liberté* that were becoming outdated. The company needed at least one ship that would enable it to compete with Cunard Line, its main Atlantic rival who, it was rumoured, was planning a new liner of about 75,000 gross tons, and the United States Lines who had recently introduced their flagship *United States*. Initially French Line posed an idea for two 35,000-ton sisters, but with morale at a low across the whole of France due to an ongoing war with Algeria, Charles de Gaulle, on behalf of the French Government, stepped in to suggest that a single ocean liner of greater size would improve national pride. The Government would loan the equivalent of $14 million to the French Line, a sum that caused much controversy amongst some ministers who were against spending such a high figure out of public funds.

Against their wishes, Hull G19 was laid down on 7 September 1957 at the Saint Nazaire Yard of Chantiers de l'Atlantique, the birthplace of France's previous superliner, *Normandie*. In fact, she was constructed on the very same slipway. But as far as construction techniques were concerned, that was where the similarity ended, for she was assembled using a method of prefabrication, a practise employed in all passenger ship construction today. Large sections, some weighing up to 50 tons, were prefabricated in towns and cities all over France – Paris, Orleans, Le Havre and Lyon, for instance – and transported to

Reflections of a new superliner. SS *France* fitting out at the Chantiers de l'Atlantique Shipyard.

Saint Nazaire for piecing together on the slipway. The liner's hull incorporated a double bottom in which up to 8,000 tons of fuel, sufficient for the round-voyage to New York, would be stored. Her fully welded hull acted as a weight-saver, as did the 1,600 tons of aluminium used for making the deckhouses on her superstructure, an idea inspired, no doubt, by the use of the same light metal in constructing upper areas of the *United States'* superstructure a few years earlier.

France's superstructure was topped off by a pair of towering funnels complete with protruding wings designed to direct exhaust fumes away from her decks into the wind of her slipstream. These impressive structures became her trademark throughout her double life. *France* was launched on 11 May 1960 by Madame Yvonne de Gaulle, wife of the French President who afterwards gave a resounding speech assuring everybody present that the Blue Riband would surely be within reach of the nation's great new liner. Yet, despite her weight-saving features and powerful engines that produced a speed of 35.21 knots on her trials, *France* never did outpace the *United States*.

Beneath the waterline impressive statistics prevailed. Her rudder weighed in at 74 tons and her four propeller shafts 53 tons each, their length of over 18m making them the longest on any ship. Her beautifully curved stem culminated in a bulbous bow. And then there was her length: 316m (1,035ft) overall, a clear 2m longer than the *Queen Elizabeth* of 1940. Although measuring 66,343 gross tons on completion, some 17,000 tons less than the Cunarder, she was the longest liner in the world.

SS *France* sailed out of Le Havre on her maiden voyage to New York on 3 February 1961. She sailed the Atlantic for thirteen years, fighting off rivalry from the airlines, and from 1966 survived financially with the help of winter cruises in warmer climes while the northern oceans were at their cruellest. She was designed as a two-class ship, an arrangement that she always retained on line voyages, but when cruising her accommodation was opened out to all her passengers. The two swimming pools were both indoors, the first class pool located deep down in her hull and the tourist class pool covered by a permanently-fixed glass dome, both designs being ideal for the Atlantic run but by no means suitable for warm-weather cruising.

In her early years, every public room was designated as either 'First' or 'Tourist', but following a few seasons of single-class cruising, it was evident that titles such as 'First Class Dining Room' or 'Tourist Class Lounge' were inappropriate, not least confusing, and thereafter every public area was allocated its own French-themed name. The First Class Grand Salon, with its raised centre ceiling over a mosaic dance floor, became Salon Fontainebleu, and the First Class Smoking Room, a beautiful area two decks high with a raised centrepiece flanked by columns, was renamed Salon Riviera, for example. Much importance was given to the design of her two dining rooms – in keeping with French gastronomic custom – and both extended over two decks. The First Class Dining Room, or Salle à Manger Chambord, was a stunning area, the centre of which featured a 5.5m-high circular dome, whilst a grand staircase allowed diners to enter in style from the deck above. The Tourist Class Lounge, or Salle à Manger Versailles, served 826 passengers of whom some were seated on its upper level.

In 1972 the French Line placed *France* on her first round-the-world cruise, her size necessitating her to pass around the most southerly tips of the African and South American continents rather than through the Suez and Panama Canals for which she was too wide. As if to emphasise the point, the disastrous finale of the legendary *Queen Elizabeth* that year, in the guise of *Seawise University*, Hong Kong, left *France* as the world's largest liner in terms of tonnage as well as length. But by this time her transatlantic schedules were under increasing pressure from the jet airliner, a fact that made the indisputable success of *France*'s new Atlantic running-mate for five seasons, Cunard's *Queen Elizabeth 2*, all the more remarkable.

Cunard Line's plan for *Q3*, the 75,000-ton liner of which rumours had reached the French Line offices, met opposition from several directions. Just as certain French ministers had opposed their Government's subsidy towards the building of the *France*, a grant of £18 million for the new British liner was also deemed generous considering the increasing numbers of transatlantic travellers who were showing a preference for air travel and the spiralling running costs of large liners. The plan was amended to the construction of a slightly smaller ship that would be capable of transiting the Suez and Panama Canals and operating as a cruise ship.

In a press release issued later, at the time of the ship's launch, Sir Basil Smallpiece, chairman of Cunard, explained the reasons for the change of direction:

> It is no secret that the major reappraisal of the passenger shipping market which Cunard instituted did not yield its findings until nearly two years after the decision had been taken to build; and what was originally conceived as an up-to-date version of the Queens became, in light of subsequent findings, something very different. From being a multi-class transport on the North Atlantic run, she has become an ocean-going hotel with transport added, partly for those whose holidays include a transatlantic crossing and partly for a worldwide cruising market. These basic changes called for a high degree of adaptability on behalf of the builders, John Brown & Co. More than any other passenger ship afloat, the new liner is tailored to long term market calculations, with a built-in flexibility that will enable her to cope with all foreseeable changes in passenger traffic during her life.

Yet he later added: 'It used to be a convention of the great liners that internally they should look as little like ships as possible and convey the illusion of grand hotels. The design philosophy of the new liner is the opposite. Not merely acknowledged but actively exploited is the fact she is a ship, and a very big ship at that.'

The amended project was named *Q4* and her keel was laid at the Clydebank yard on 5 July 1965. Steadily and surely Yard No.736 grew on the same slipway that saw the birth of the two Queens. Like her French counterpart, the new ship was assembled in prefabricated sections and her Upper Boat, Observation and Sports Decks were constructed entirely of aluminium – around 1,100 tons in all – a weight-saver that would reduce her otherwise deep draught by more than 2m. The use of alloy was further extended to the liner's interiors, not only for use in window surrounds and handrails but in furnishings and decor. The maritime world had entered a new era of safety consciousness and fire regulations disallowed the use of the real wood so evident in older liners. *Q4*'s corridors would be veneered in cedar and unprecedented amounts of Formica laminate would feature throughout her public areas and cabins, cleverly textured to overcome the appearance of cold plastic. But these would be part of her fitting out process after her launch, and for now *Q4* was still in early days of construction, her design entrusted to the hands of the experienced James Gardner and Dan Wallace and a team of six naval architects who had worked for over five years on everything to do with the ship apart from her machinery and decor.

In a press release issued shortly before the liner's launch, John Rennie, managing director of John Brown & Co., put into perspective the immense amount of work involved in the pre-planning:

> Every piece of this great operation has been planned with great detail. We established a special planning department with a computer to help us in the countless intricate calculations. About 40,000 men-weeks were spent on planning, design, drawing and calculating. To give an example of all the work involved in planning, there will be 100,000 pipes of all shapes and sizes and materials. All have to be drawn, fashioned and fitted.

The 1960s in Britain was a period of great optimism as the nation finally rid itself of the shackles of post-war austerity. In 1966 the England football team led the world, a feat never since repeated, yet that same year witnessed a national seamen's strike that brought Britain's docks to a standstill and cost Cunard some £4 million. The working public had more spare cash in their pockets than they had ever known, yet disputes and strikes were becoming the vogue, a problem that soon arrived at John Brown's shipyard. Construction of *Q4* was already six months behind in 1966. Just a year later Cunard made the shock announcement that both *Queen Mary* and *Queen Elizabeth* would be retired from service by the end of 1968. So, on *Q4*'s delivery Cunard would be left with a single ship on transatlantic duty and, although much smaller than the legendary pair, she would be Britain's biggest liner.

On 20 September 1967 HM Queen Elizabeth II stepped forward to end all speculation concerning the new liner's name. Cutting the cord with the same pair of gold scissors used by her mother to launch the *Queen Elizabeth* and her grandmother to launch the *Queen Mary*, she announced: 'I name this ship Queen Elizabeth the Second. May God bless all who sail on her'. By appearing to name the ship after herself, the Queen unwittingly caused much debate as to whether the liner was honouring her as British sovereign or simply that she was Cunard's second *Queen Elizabeth*, and the debate lasted until the end of the ship's working life. Cunard quickly modified the name so that it featured the numeral '2', and even before her completion the liner became widely known as the *QE2*.

Three thousand shipyard workers were employed in her fitting out during which several amendments to her intended interior design were made. The six-month delay had enabled Cunard to take a step back and review *QE2*'s requirements, opting to reduce the planned three classes of passenger accommodation to just two, making the ship more easily adaptable for cruising. Dennis Lennon was a joint design co-ordinator responsible for many of the public areas. He highlighted how even the smallest alteration seemed to grow out of all proportions: 'We made fourteen sample cabins for the 700 to be built. Sometimes there would be anything up to thirty people in a cabin arguing about where a light switch should go. Anything they wanted to add had to be multiplied by 700.'

Progress towards the *QE2's* completion remained subject to the occasional industrial dispute and even unconfirmed reports of petty thieving. Her builders' financial position was becoming ever more perilous and amid concerns that they would go into receivership, John Brown & Co. merged with four other Clydebank yards – Charles Connell & Co., Alexander Stephen & Sons, Yarrow & Co. and Fairfield (Glasgow) – to form Upper Clyde Shipbuilders in February 1968. So *QE2* was effectively brought to completion by a different concern from that which began her. By the end of that year she was at last finished and set off on a cruise to the Canary Isles, her acceptance trials. *Queen Elizabeth 2*, as we should refer to her from time to time, was equipped with two sets of double-reduction geared Brown-PAMETRADA steam turbines manufactured by her builders, three Foster-Wheeler ESD II-type boilers built under license, also by John Brown & Co., and two six-bladed fixed pitch propellers that drove her to an impressive maximum trials speed of 32.66 knots.

It was the turbines that let her down first. She would remain vulnerable to turbine and other mechanical problems throughout the first half of her commercial life, but her turbine faults and the fact that her interior was oceans away from a satisfactory completion left Cunard with no option but to refuse delivery of the liner. The British media were quick to hurl criticism at their nation's new flagship, but after three further sets of trials and a shakedown cruise, *QE2* made an eight-day fare-paying cruise before sailing proudly out of Southampton to begin her first transatlantic crossing and a famous career. She was given a huge welcome at New York and achieved a further twelve round-voyages during her initial year, operating at 80 per cent capacity. But *Queen Elizabeth 2* was more than just a transatlantic liner; she was also a cruise ship, the biggest and most spacious cruise ship afloat.

Her beautiful lines, so easy on the eye, were a great tribute to Wallace, Gardner and their teams, whilst her 21.36m-tall funnel, painted white and black, a surprising diversion from Cunard's traditional red and black, created much interest. As James Gardner had explained to the press during *QE2's* construction:

> The distinctive feature of sailing ships was their rig. New ships are sealed, air-conditioned boxes and their distinctive features are their breathing points, ducts and vents. Rationalization of services has cleared away the odd obstructions associated with ships' decks and all the major ducts erupt at one point, just forward of centre. In place of the traditional cylindrical smoke stacks we have an unexpected form which appears at first sight to be notional. Used air is directed upwards to help keep the boiler gases up and away from the decks. The forward mast is also a vent, designed to echo that of the central superstructure and complement it.

Making herself at home. Displaying her original white-painted funnel, the newly completed *QE2* is berthed alongside the QEII Terminal at Southampton.

Cunard had originally planned a ship of 58,000 gross tons, but thanks to ongoing alterations during her construction *Queen Elizabeth 2* measured 65,863 gross tons by the time she entered service. With dimensions of 293.53m in length and a width of 32.09m, she cost, in total, £29,091,000 to build. Designer Dennis Lennon had, of course, referred earlier to the ship's 700 cabins, but she set sail with 978 that could accommodate up to 2,005 passengers on transatlantic service, although her capacity was limited to a more comfortable 1,400 passengers whilst cruising. Lennon's work included the designing of the 500-seat Columbia (first class) Restaurant, the Grill Room (of 100 seats) and Midships Bar on Quarter Deck; the Britannia (tourist class) Restaurant (of 815 seats) on Upper Deck and the uniquely circular Midships Lobby (then first class) on Two Deck. The forever beautiful Queen's Room (Quarter Deck) and the two-storey appropriately-named Double Room, an entertainment venue, remained memorable public areas for years to come. A 'swinging sixties' theme was prevalent throughout, this musical era being well represented by the ship's disco, the 736 Club, named after her builders' number.

Below deck, *QE2*'s machinery weighed just one-third of the combined weight of the heavy turbines and boilers carried by the old *Queen Mary*, saving up to £1 million per annum in fuel costs compared with her pre-war predecessor. But, as the 1970s progressed, the worldwide cost of oil escalated to an unprecedented level, and without the *QE2*'s pre-planned adaptability for cruising she would surely have struggled to make ends meet.

In 1973 the price of oil spiralled from US$35 to US$95 per barrel. This proved to be the end of the road for *QE2*'s French partner, for on realising that the cost of keeping the *France* in service would increase by no less than $10 million a year, the French Government decided the money would be better spent on subsidising the supersonic airliner Concorde that was under development at the time. Without this money French Line would be unable to operate and it was announced that *France* would be withdrawn from service on 25 October 1974. Like *QE2*, *France* had been designed so that her two-class accommodation was arranged horizontally rather than vertically (the higher the class, the higher the deck number), but the shipyard delays incurred by the Cunarder had, in the long run, proved advantageous for her, as her owners were presented with the time to suitably modify her for cruising. The *France* was not so easily convertible for cruising and, furthermore, being too wide for the Suez and Panama Canals made her less suitable for deployment worldwide.

On hearing the news the crew of *France* voted to take strike action and the liner was anchored in the entrance to her home port of Le Havre, her passengers having to be ferried ashore. The hijacking failed but *France* as a transatlantic liner was finished, and once the dispute had died down the liner found herself moored at Le Havre's most distant quayside – the Quai de l'oubli – the Pier of the forgotten. The sad sight of the mothballed giant, her interiors still filled with furniture, brought a tear to many a local's eye – a further explanation for those emotional Francophile welcomes afforded her in her later years.

In 1977 hopes were raised by millionaire Akram Ojjeh from Saudi Arabia who proposed to employ her as a floating museum for his collection of French antique furniture in which he had invested $15 million. Not only that, the ship would act as a museum of French civilisation and a casino and hotel, being moored off Florida's Daytona Beach. A price in excess of $20 million was paid, but his grand ideas were never realised and *France* was placed back on the market.

Over the next months there were rumours of bids from the former Soviet Union and from China until, after four full years of lay-up, the liner was purchased for the sum of $18 million and, moreover, for employment as a passenger ship. Her saviour was Knut Kloster, founder of Oslo-based Norwegian Cruise Line. A brand-new era of Caribbean cruising had already seen the delivery of a number of purpose-built cruise vessels, but they were of modest size in comparison with the former French flagship. Kloster was an ambitious man and his company was already proving to be one of the most successful on the Caribbean circuit. His new acquisition, that had 377 transatlantic crossings and ninety-three cruises behind her, would cost a massive $80 million to convert into the world's largest cruise ship, the first the Americans could rightly call 'a destination in herself'.

In August 1979 *France* was towed to the Hapag-Lloyd Shipyards in Bremerhaven where they would carry out the huge renovation project. Eight months later, her deck space expanded and ready to catch the Caribbean sun, she was unveiled as the *Norway*. Registered in Oslo, she raised the flag of the United Nations as a representation of her multi-national crew alongside the colours of Norway. Throughout her cruising career she would remain the only ship allowed to fly the UN flag. Kloster's team of designers had shown great sympathy towards her original external profile, her huge uniquely-shaped funnels thankfully surviving an unnecessary proposal for their replacement. Norwegian Cruise Line concentrated on her aft areas by extending her open deck space, creating a stepped formation of lido decks so vast that it actually cantilevered over her stern quarters. This area housed a new outdoor pool and buffet restaurant, idyllic, as I can personally vouch, for idling away the morning hours over breakfast. A second swimming pool and a squash court were inserted high up between her funnels, flanked on both sides by her new name emblazoned in expansive gold lettering.

As speed was no more an important issue – *Norway* would essentially be an island-hopping cruise vessel from then on – her forward engine room was removed together with two of her four propellers. With her aft engine room modified to operate the two remaining screws at around a quarter of her previous power, she would cruise at 18 knots, thus reducing her fuel bill by over two-thirds. The addition of bow and stern thrusters (a real godsend for modern cruise liners) enabled the mighty vessel to turn in her own length, more often than not without the assistance of tugs. Forward of the bridge, towards the fo'c's'le, were positioned a pair of double-decker tenders, built in Norway by Holen Mekaniske Verksted. With room for 400 passengers and a speed of 11 knots, each would be employed to transfer passengers between ship and shore on occasions when *Norway*'s 10m draught prohibited her from tying up at the dockside. Held in place by two giant davits, the tenders – named *Little Norway I* and *Little Norway II* – were themselves registered as ships, thus making the *Norway* the world's only passenger vessel to carry other ships.

Internal changes made by Kloster ensured that *Norway* would fit comfortably within the American vacation market, with some subtle Parisian touches in recognition of her former life. The work was carried out under the direction of marine architect Tage Wandborg and New York interior designer Angelo Donghia who completely redesigned all the tourist class public areas, including the port and starboard promenades on the main tourist class deck (Pont Promenade) which were fitted out with new prefabricated junior suites.

One deck above, the promenades on the first class deck (Pont Veranda) were retained and transformed into the main thoroughfare used by passengers circulating the ship's principal amenities deck. Renamed Fifth Avenue on the port side and Champs-Elysées on the starboard (maintaining the American/Parisian theme) the sheen of their mosaic-tiled flooring and stylish wrought-iron street lamps made them a veritable magnet to which I was drawn time and again. Towards the aft end, before exiting outside to the 'Great Outdoors Restaurant', I came across Club Internationale, a classy cocktail lounge-cum-small ballroom, converted from the original two-storey first class smoking room (Salon Riviera). In typical French style, great importance had been attached to the ship's dining facilities in her transatlantic days, and this was continued, no doubt with her food-conscious American clientele in mind. The first class restaurant (Chambord), remained

A unique pairing. *Little Norway I* and *Little Norway II* in their positions overlooking the bows of their 'mother ship', the *Norway*.

Ocean boulevard. *Norway*'s starboard promenade Champs-Elysées, its mosaic flooring shining in the morning sunlight.

Norway's Club Internationale featured daily entertainment in classy surroundings.

almost unaltered, a brilliantly astute decision by NCL who even retained its sweeping 'grand entry' staircase and original light wood panelling. The Windward Dining Room, as it was now known, was not, sadly, my allocated formal eating place, but I managed to sneak in for an in-port luncheon which allowed me to fully appreciate the wonderful ambience of this circular room. More work, however, was undertaken on the tourist class restaurant (Versailles), with the addition of an aluminium chandelier and a new spiral staircase that linked the room's two levels. I found the Leeward Dining Room, as it was renamed, quite crowded, a shortage of elbow room for the dining staff causing the odd spillage or breakage. Yet the room seated 604 on its lower level and 146 on the balcony, compared with 552 in the Windward, necessitating two dinner sittings nightly in both dining rooms.

Norway emerged from the Bremerhaven shipyard sporting a royal blue hull and with her funnels painted in dark and light blues and white. She made a celebratory visit to her port of registry, Oslo, before moving on to Southampton for the commencement of a transatlantic voyage to New York, then on to her new home port of Miami. Settling into a regular pattern of seven-night cruises, she was an instant Caribbean success. Now measuring 69,529 gross tons, she was the world's largest cruise liner, taking over top place from *QE2*, which was yet to be employed in full-time cruising. Such was her passenger capacity – almost 2,000 – it would have taken four of her Caribbean companions to accommodate the same number. Consequently she was infinitely more economical

Norway's wonderfully circular Windward Dining Room was little changed from its Francophile days as the first class restaurant Chambord.

to run. Not only did her sheer size cause the cruise industry to sit up and take note, but the positioning of her onboard shops, cafés and principal entertainment facilities alongside her 'main street' promenades proved an inspiration to subsequent passenger ship designers, especially those of Royal Caribbean's twenty-first-century floating cities.

Keen to build on her success, Norwegian Cruise Line upgraded its flagship at Bremerhaven in 1982 and again two years later. On the second occasion the company opted to combine the refit with a series of Norwegian fjord and Baltic capital city cruises, a somewhat strange decision considering her continuing popularity on the other side of the Atlantic. Furthermore, the response to *Norway's* two transatlantic repositioning voyages was so disappointing that she was not to be sent to Europe again for twelve years. But competition was beginning to close in on her home patch as the Carnival Corporation, still very much in embryonic form, and Royal Caribbean introduced their latest breed of purpose-built cruise vessels. *Norway's* standing as the world's biggest cruise ship was under no serious threat until 1988 when Royal Caribbean's 73,192-gross-ton *Sovereign of the Seas* arrived, ironically, from the same Saint Nazaire shipyard that had built *France/Norway*, to take over that accolade. This innovative ship will be profiled in the next chapter.

The innovative Roman Spa and Health Centre used to be the indoor swimming pool when *Norway* was sailing as the *France*.

It was time for Kloster's company to take serious action and *Norway* was despatched on a non-commercial voyage to Bremerhaven for further expansion. Still the world's longest passenger vessel, she could hardly be expanded lengthways, so two new decks were added to her superstructure, a full-length Sky Deck and a much shorter Sun Deck that was erected around her forward funnel. In all, 135 new suites were installed, including a pair of Owner's Suites, then the largest afloat. Another aspect of the refurbishment was the alteration of the ship's indoor swimming pool into a Roman spa and health centre, a forerunner of one of the most essential amenities to be found on today's cruise ships. On her return to the Caribbean some ship lovers expressed their indignation at *Norway*'s extra height, claiming that the new look spoiled her former majesty, but NCL's decision to enlarge her proved a shrewd move, for not only did the new suites bring in valuable additional revenue but *Norway* reclaimed her crown as the largest cruise ship with a revised gross tonnage of 76,049.

As I boarded *Norway* at Southampton's QEII Terminal that late August afternoon, my eyes were drawn to her new funnel colours. The shades of blue and white had been replaced by a royal blue (to match her hull) on which the letters 'NCL' in gold with gold surround were superimposed. But the French cared not a notch about the ship's

The locals throng Le Havre's quayside to catch a close-up view of their nation's former flagship.

change of look as they flocked in their hundreds to the Le Havre quaysides to greet their old friend. Around 6,000 fortunate soles had been selected to go on board while others simply stood and stared in awe. I strolled into Le Havre town and bought a copy of the *Normandie* newspaper. '*Il est avec nous!*' exclaimed the main headline: 'It (the *Norway/ France*) is with us.' Alongside, as the second headline, was news of Princess Diana's tragic death. Suddenly I felt very humble.

Our departure from Le Havre was more spectacular than I could have ever imagined. Memories of that special evening are still sharp in my mind today. At 10 p.m. exactly *Norway* eased away from her berth and, in tandem with *Maasdam*, made a grand floodlit exit into 'la Manche'. Fireworks rose dramatically from the French shoreline that was packed with an estimated 100,000 waving, cheering people. *Norway* saluted them with her reverberatingly deep whistle until we could see them no more.

I had a great affection for the *Norway*, as did New York maritime historian and writer John Maxtone-Graham whose small but fascinating onboard museum of artefacts from the ship's days as the *France* would allow him the perfect excuse to pay regular visits to her in order to keep the exhibits clean and polished. Yet *Norway* did have her shortcomings. She had already been eclipsed, in terms of gross tonnage, by the *Sun Princess* (77,441 gross tons), delivered to Princess Cruises in 1995, and ships of even greater proportions were not far over the horizon. Unlike these new floating resorts, *Norway* could not offer the added luxury of balconied cabins, her principal two-storey theatre seated only 766 and the acclaimed Club International just 160 passengers. Considering her overall size and new passenger capacity of 2,565, such factors were hindering her bid to hold on to her reputation as the Caribbean's most popular cruise ship. On 25 May 2003 *Norway* suffered serious damage from a boiler explosion shortly after docking at Miami, taking the lives of seven crew members. NCL sent her to Bremerhaven one last time. Construction of a new boiler was out of the question, although parts could be found to repair her.

Then, in March 2004, NCL announced '*France* will never sail again'. There followed a difficult, protracted death of the old liner; a slow, sad journey to an Alang beach in India where dismantling did not even begin until the end of 2006. Renamed *Blue Lady*, a title that encapsulated her elegant charm right to the end, she was the victim of inaccurate reports, rumours and even court cases until she was eventually turned into a metallic mountain in favour of twenty-first-century white-painted floating resorts, the very ships that her own interior designs had inspired.

I bade farewell to *Queen Elizabeth 2* in July 2008, just less than four months prior to her withdrawal from Cunard service. In a similar manner to *Norway*'s retirement, the announcement of her termination as an active cruise liner was met with great sadness and no small amount of surprise by ship lovers, especially as she had received more facelifts than most middle-aged international celebrities. As I write these words she is amid her final voyage, en route for the sun of Dubai, and books and DVDs of her lifetime are flooding onto the shelves and the internet. But suffice to say that her near-forty-year career was as eventful as a blockbuster movie as she moved from villain to hero – villain for her mechanical unreliability, before Cunard realised they could completely win over her critics by replacing those troublesome engines, and hero thanks to her participation in the rescue of 500 passengers from the burning French liner *Antilles* (a former fleet-mate of the *France*) in 1971 and her brave role in the Falklands War of 1982. Just like the *Canberra*, her 'sister' troop transporter during the conflict, QE2 quickly found herself elevated to legendary status within shipping circles and remained a household name thereafter. She conveyed 3,000 troops and 650 volunteer crew to the South Atlantic, having been prepared for war in just one week.

By contrast, her peacetime refit in Southampton's King George V Drydock cost £7 million and took nine weeks to complete, allowing Cunard time to alter her funnel colours to the company's traditional red with two black bands and a black top, a move that delighted all her admirers who, however, were not as happy with the changing of her hull colour to light pebble grey. Among the interior alterations carried out during the refit was the conversion of her indoor swimming pool into the 'World's First Floating Spa', which featured three large whirlpools, installed in place of the Turkish Baths, saunas and a climate-controlled gymnasium. Incidentally, another 'first' for QE2 was the installation of the first seagoing branch of Harrods in 1984, a sure winner as far as her affluent American female passengers were concerned.

BUILDING THE BIGGEST

The steam turbine powerplant that had driven *QE2* since her completion had been a constant headache for Cunard. Not only had it proved unreliable but by the end of the 1980s was consuming 600 tons of fuel a day and spare parts were becoming increasingly difficult to acquire. Should her owners replace their flagship or re-engine her? The second option was deemed more practicable as this would be far less expensive and be considerably quicker than building a replacement ship. A new, more efficient diesel-electric powerplant would enable *QE2* to continue for at least a further twenty years, a calculation that, as we now know, was to prove accurate.

On 27 October 1986 *Queen Elizabeth 2* entered the Bremerhaven shipyard of Lloyd Werft, who would be undertaking the biggest merchant ship conversion ever known. In all, the work took 179 days, cost in excess of £100 million, over three times *QE2*'s original building cost, but presented Cunard with a virtually brand new ship. Nine German MAN nine-cylinder, medium-speed diesel engines were installed, each of which would drive a GEC generator. This electrical plant would, in turn, drive two main 400-ton propulsion motors (as well as powering the ship's hotel and auxiliary services through transformers), one on each propeller shaft. Notably, her normal service speed of 28.5 knots could be maintained by using seven of the nine sets, so it was easy to appreciate how much more efficient and less expensive the new diesel-electric propulsion would be to run. Two five-blade 5.8m diameter variable pitch propellers were also fitted. They could easily be reversed by themselves rather than by the astern turbines that formed part of the old powerplant. Unfortunately the new screws did have a few problems, and in 1988 they were replaced by a slightly larger 6.1m set. Thereafter *QE2* was capable of moving in reverse at up to 19 knots, faster than any other ocean passenger ship afloat.

Externally, *QE2* was given a facelift that, according to popular opinion, enhanced her already attractive lines. Her hull had already been repainted black, much to the relief of her enthusiasts, and she was given a replacement funnel which, at 21.2m, was the same height as the previous structure, but clearly wider, giving her a sturdier and more powerful appearance. Internally, no less than £25 million was allocated to improvements to the 'hotel' aspect of *QE2*. Eight category 'A' penthouse suites were added to a block of ten suites installed in 1972, forming a continuous row of suites between funnel and mast. The Queen's Grill, an exclusive dining area for occupiers of higher-graded accommodation, was enlarged. This lovely room, together with its galley and the adjacent Queen's Grill Lounge, had also been installed during *QE2*'s 1972 refit, replacing the original Coffee Shop, Juke Box Room and 736 Nightclub that belonged to the 'swinging sixties' age.

Larger public areas were similarly improved and altered. On Upper Deck the original tourist class Britannia Restaurant that in 1977 had been redesigned by Dennis Lennon and renamed the Tables of the World, received a £2 million refurbishment and was now known as the Mauretania Restaurant. The two-level Double Room, the ship's principal show and cabaret venue, became the Grand Lounge and was given a new double-sided stairway linking both levels. The shopping arcade on the upper level was upgraded and extended aft by some 16m. On Seven Deck a huge new laundry room was constructed.

Back at her birthplace. *Queen Elizabeth 2*, her funnel colours amended to red and black, makes a rare return to the Clyde.

On 28 April 1987 the revitalised *Queen Elizabeth 2* arrived back at Southampton to prepare for her 'second maiden voyage'. She had a day to spare, so 500 local schoolchildren were invited to a huge onboard children's party, with one very special guest. When HRH The Princess of Wales appeared at the top of the Grand Lounge stairs the children clapped and cheered. A memorable day for everyone involved but, unfortunately, *QE2*'s first voyage as a diesel-driven liner was to be remembered for the wrong reasons. Work on some areas of her accommodation was still incomplete, while many new crew members had just joined the ship, meaning the high standard of service that her passengers had a right to expect failed to be attained. But this was a temporary blip as the famous ship's reputation continued to grow amid an expanding worldwide cruise market.

Further refits were to follow as Cunard kept their only transatlantic vessel in line with her discerning cruise and line-voyage passengers' growing expectations. In 1992 her health spa facilities on Six Deck were revamped. The facility's swimming pool was replaced by saunas, a steam room, whirlpool spa bath, inhalation room and a Thalassotherapy pool. Her other indoor pool on Seven Deck was retained but surrounded by a new fitness suite. 1994 saw yet another refit, carried out, like her previous overhaul, by Blohm & Voss in Hamburg. A workforce of 2,000 plus 400 crew were involved in the thirty-two-day contract that cost the equivalent of £1 million per day.

Added in the late '90s, a row of luxury suites extend from funnel to mast on *QE2*.

Many of the refurbishments undertaken in that refit as well as the spa improvements of 1992 were still very much in evidence even in the final months of *QE2*'s seagoing career. These included the relocation of the Columbia Restaurant (originally first class) from Quarter Deck to Upper Deck to become the Caronia Restaurant, the formal dining room allocated to passengers in medium-priced accommodation. Taking the place of this, on Quarter Deck, was the Mauretania Restaurant, used by passengers booked in lower-grade cabins. New rooms were the Crystal Bar and Golden Lion Pub, reminiscent of a traditional English public house, whilst the Midships Lounge was modified as the Chart Room, a small friendly lounge. A new Yacht Club, constructed at the aft end of Quarter Deck, replaced a swimming pool area enclosed by a magradome glass roof. Subsequent refits at Southampton in 1996 and Bremerhaven in 1999 (costing £30 million) concentrated on the development of luxury suites. 'Queen Mary' and 'Queen Elizabeth' suites were enlarged into Grand Suites and the ship's old Radio Room and two senior crew member's cabins were converted into the Caledonia Suite and Aquitania Suite respectively. Additionally, the Caronia Restaurant was redesigned in a style of an English country house. Then, in 2004, another refit included the creation of a new Funnel Bar, located just behind the ship's funnel.

From 1994 *QE2*'s gross tonnage had increased to 70,327.

Once news of *Queen Elizabeth 2*'s retirement to Dubai had become public knowledge it seemed that everyone wanted to grab an opportunity to sail on her during those final few months. Her 2008 farewell season was a sell out (her final, one-way cruise to Dubai reputedly sold out in just thirty-six minutes) and included a northern European cruise that represented my personal 'goodbye' to the great lady. Boat Drill in the Caronia Restaurant revealed a wide age range of passengers, from mature and experienced seafarers to teenage girls and young children. I was sharing a low-grade outside cabin on Five Deck, a room that was evocative of *QE2*'s transatlantic liner days rather than a stateroom of a modern-day cruise ship. The walk-in closet was spacious enough, so was the bathroom (although, thanks to a sticking shower tap, the ship was under frequent threat from flooding!), but the stowed away optional third bunk would have allowed notably restricted headroom for the occupier of the bed below, had it been in use. But, hey, this was the *QE2*, and as long as there was enough elbow room for putting on your evening attire for Royal Night, who cared?

Passengers strutted their stuff – often in the wrong directions, confused by the internal layout – as *QE2* oozed with night-time glamour. The timeless Queen's ballroom, surely capable of hosting any televised dance contest, the Grand Lounge with its West End shows and cabaret entertainers, and the cosy Chart Room that featured a grand piano from the old *Queen Mary*, where a young lady harpist from Wales would charm us with her soothing music; all were exquisite. But regrettably there was no disguising the fact that time was catching up with the great ship, and she did have her faults: the weathering of window surrounds, for example, the limited seating on the upper level of the Grand Lounge due to her souvenir shops whose quality hardly did justice to the ship's heritage, and those cramped lower-grade cabins.

Just after 7.15 p.m. on Tuesday 11 November 2008, with tens of thousands of onlookers crowding Southampton's waterfront, *Queen Elizabeth 2* left her home port for the final time. As one last reminder of her ability to reverse at speed, the thirty-nine-year-old ship moved astern from the QEII Terminal, coming to a halt in front of Mayflower Park to salute her waving and cheering fans. Then, amid a huge firework display and a tumultuous din of hooters and horns to which she replied with several thunderous blasts of her famous whistle, *QE2* moved slowly down the Solent and into the night. It was all very emotional, too much, perhaps, for *QE2* herself, for she had briefly ran aground on a Solent sandbank early that morning, whilst during that final fifteen-night voyage to Dubai leaks appeared in a few of her cabins and some of her passenger lifts jammed. She was to be treated, though, to a Dubai welcome just as impressive as her departure from Southampton. Accompanied by numerous small craft and the Royal Navy frigate HMS *Lancaster*, she tied up at the port of Mina Rashid. Captain Ian McNaught gave his final order, 'Finish with engines', and her 1,800 passengers thronged her decks to enjoy the now-compulsory fireworks display, and then spent their last night aboard. Veteran British entertainer Des O'Connor was given the honour of being the last act to perform onboard.

On 27 November her passengers departed, and *QE2* ceased to be a Cunarder as her handover to Nakheel, part of the Dubai World organisation and a key player in the massive development of Dubai in the twenty-first century, was formally completed.

1 The ship that refused to be launched. Brunel's *Great Eastern* at speed during her early passenger-carrying days.

2 SS *Great Britain* back at her birthplace. She is now a familiar Bristol landmark and is open to the public daily.

3 Known as the 'Queen of the Ocean', *Oceanic* was the world's biggest liner at the start of the twentieth century.

4 Show off! Sir Charles Parsons demonstrates the abilities of *Turbinia* at the Spithead Review.

Recalling transatlantic times, the poster displayed in the author's cabin on *Norway* in 1997.

6 Made in France. *Monarch of the Seas* under construction in her building dock at Saint-Nazaire.

7 *Oriana*'s birthplace, Meyer Werft's Papenburg shipyard, is dominated by its huge building hall.

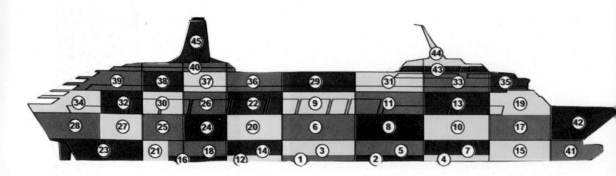

8 From 1 to 45, *Oriana* was assembled in prefabricated blocks.

9 Evoking the days of the classic two-funnelled ocean liners, *Disney Wonder* (pictured) and *Disney Magic* have met with much approval from shipping aficionados.

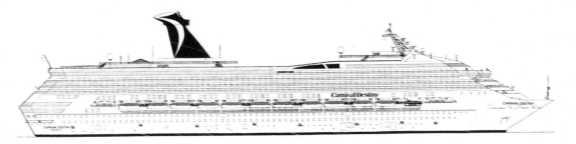

10 A builder's image of *Carnival Destiny*, the first passenger ship to top 100,000 gross tons.

11 Crossing blue waters. *Carnival Destiny*'s external profile is clearly enhanced by her angular funnel and rounded forward superstructure.

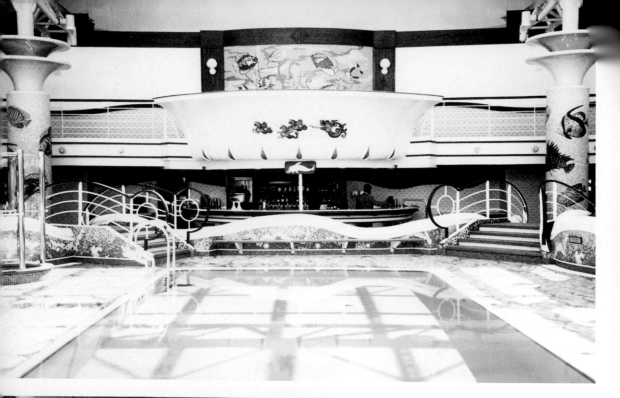

12 Splashing out. The attractively ornate covered lido on *Golden Princess*.

13 The long and winding road. The impression of length has been cleverly concealed by the designers of the Royal Promenade on *Voyager of the Seas*.

14 Scaling the heights. The rock climbing wall is now a feature on all Royal Caribbean ships.

15 *Explorer of the Seas* makes an impressive sight as she arrives at Southampton during her delivery voyage.

16 *Left:* The 154,400-gross-ton *Freedom of the Seas* fitting out at Turku. She was Royal Caribbean International's biggest ship at the time.

17 *Below left:* Newly completed, *Queen Mary 2* lies in dry dock at Saint-Nazaire. Scaffolding near her bows indicates that work on her forward thrust is still proceeding.

18 *Below:* A photo-shoot with a difference. Shipyard workers gather under the massive stern of *Independence of the Seas* just prior to her float out. The propeller of the middle fixed Azipod faces aft, unlike those on the two outer rotating units.

19 The Royal Promenade on *Independence of the Seas*, overlooked by inward-facing cabins, is even grander than on previous ships.

20 *Opposite top:* Under the direction of an Aker Yard's worker, the first keel section of *Oasis of the Seas* is swung into position.

21 *Opposite centre:* Block by block, *Oasis of the Seas* is taking shape. Note the building hall to the left, of insufficient size to accommodate the huge megaship.

Opposite bottom: May 2009, and with her massive bulbous bow already in place, work on *Allure of Seas* is well under way.

23　A modern variation. The transom stern of *Oasis of the Seas* is designed around her Aquatheatre.

26 The garden theme to a new extreme. The *Oasis*-class duo's Central Park is based upon the famous New York landmark of that name.

24 *Opposite above:* The stern of the twenty-first century. The more complex transom design of *Explorer of the Seas*, with added 'skirt'.

25 *Opposite below:* Celebrity Cruises claim to be pioneers of the 'AquaSpa', one of the largest and best-equipped spas at sea, as shown here on *Celebrity Summit*, 90,280 gross tons.

27 With the advent of jet airliner travel half a century ago, huge ocean passenger ships appeared condemned to history. But, like the novelty Carousel featured on each *Oasis* megaship, they have completed a full circle, coming around again as specialist cruise liners. How big will they one day be? We can only wait and see.

CHAPTER FIVE

A CRUISE ERA EVOLVES

It was on 19 December 1966 that it happened. A small cruise vessel set sail from Miami, bound for Nassau, an event that seemed perfectly innocuous at the time but would irrevocably change the future of world passenger shipping. Although commercial cruising was invented as long ago as the 1880s, its rebirth, following years of predominately ocean liner travel, can be traced back to that day. The *Sunward* had been built as an 8,666-gross-ton cruise ferry for service between Southampton, Vigo, Lisbon and Gibraltar. Effectively a larger version of the Scandinavian ferries built around that time, she was being operated

The world's first modern-day cruise ship, *Sunward*, pictured in her European ferry days prior to her transfer to the Caribbean.

Norwegian Cruise Line suitably promoted its early 'Wedge Ship' *Starward* under the Florida sun.

by Kloster Sunward Ferries, a new company owned by the well-known Norwegian ship-owners, the Kloster family. But the service, despite early optimism, proved unsuccessful and the *Sunward* was transferred to Florida to be run by Norwegian Caribbean Lines, a recently-formed venture co-ordinated by Knut Kloster and Ted Arison, an Israeli-American businessman who was to go on to found Carnival Cruise Lines in 1972.

Such was the response to *Sunward*'s short cruises, two slightly larger purpose-built cruise ships were ordered from Germany. *Starward* and *Skyward* both began a seven-day Caribbean cruise schedule in 1968 and were followed by the similar *Southward* in 1971. *Sunward* was sold off to French owners in 1972, but the remaining three continued to prosper on Caribbean services for some years, only to be dwarfed in 1979 by the rebuilt *Norway* which, as we saw in the previous chapter, changed the outlook of the cruise industry with her immense size.

The smaller trio managed to make their own mark thanks to their 'Wedge Ship' design with a high superstructure, foreshortened bows and machinery and a funnel positioned aft, or at least aft of centre. They imitated the modern Baltic ferry, built to accommodate maximum restaurant and entertainment space. They were a breakaway from the classic ocean liner with its centrally positioned smoke stack (or maybe two), sweeping superstructure and long fo'c's'le, but, regardless of shipping enthusiasts' opinions, they were clearly pointing the way to twenty-first-century cruise ship design.

Why the Norwegians should have had such a strong interest in the Caribbean has never been clear – apart from the lure of year-round sun, of course – but there was no denying their connections as the Nordic theme increasingly featured in the names of ships and ship-owning companies. By the 1970s Norwegian Caribbean Line had become Norwegian Cruise Line (thus retaining the initials NCL), the three-ship Royal Viking Line joined the Caribbean circuit in 1973 and Royal Caribbean Cruise Line was already well under way with its first batch of purpose-built cruise vessels. Formed in 1968 by the well-established Norwegian shipping concerns Anders Wilhelmsen & Co., I.M. Skaugen & Co. and Gotaas Larsen, Royal Caribbean set up its headquarters on the Miami seaboard and took delivery of its first ship, *Song of Norway*, two years later. Like her two sisters, *Nordic Prince* and *Sun Viking*, delivered in 1971 and 1972 respectively, the 18,417-gross-ton vessel could accommodate 724 passengers and was constructed at Helsinki by Oy Wartsila Ab ('Oy' is the abbreviation of Osakeyhtio, Finnish for 'Incorporation', and Ab stands for Aktiebolag, Swedish for 'Incorporation', so presumably both were included for bi-national purposes). Wartsila, incidentally, took its name from a port to the east, in Karelia, to be precise, where a sawmill was established in 1834 and a blast furnace for the processing of local iron ore was constructed later on. The company was formed by Wilhelm Wahlforss in 1938, but Finnish shipbuilding can be traced back to 1738 when a small yard for the building of wooden sailing ships was opened at Turku, and to 1865 when a shipyard in Helsinki was founded. These two locations remain the prime shipbuilding centres of Finland today, but much water has passed under the proverbial bridge along the way.

The war years virtually destroyed the Finnish yards, but they rose from the rubble to build a series of ice-breakers for the former Soviet Union, Baltic ferries and, of course, cruise vessels, including Royal Viking Line's handsome trio. By then Royal Caribbean's three sisters had added a new dimension to cruise ship design with their distinctive Viking Crown Lounge, an observation area attached to the ship's funnel that would become the company's recognisable trademark over the years. In 1982 Royal Caribbean returned to Wartsila to take delivery of *Song of America*, which, at 37,584 gross tons, was a larger example of the initial class, sporting an observation lounge that actually completely encircled her funnel.

In 1984 P&O Princess accepted its first Finnish-built cruise ship in the pleasing shape of *Royal Princess*, larger again at 44,348 gross tons and featuring a cruise industry 'first' – a number of balconied cabins. This innovation proved immensely popular with her clientele and began a new era of multi-balconied ships that, in some cases, would struggle with their outward appearances. Wartsila, however, had a struggle of a different kind, for as glasnost approached, the Soviet Union, a market upon which they had so much relied, was no longer wanting new ships. In 1989 the shipyard group went bankrupt, but shipbuilding executive Martin Saarinkangas was on hand to create Masa Yard just a year later, taking over the Wartsila Yard. Another year on and another change as Masa Yard became part of Norway's Kvaerner group to become Kvaerner Masa-Yards. And the name changes have refused to cease. In 2005 Kvaerner Masa-Yards merged with Aker Finnyards, owner of two shipyards at Rauma, to officially become Aker Yard Oy in June 2006, only to become STX Finland Oy in November 2008 following its acquisition by the massive Korean-based international industrial group STX. Under a new banner, STX

The crowning glory. The Viking Crown Lounge completely encircled the funnel on the Royal Caribbean's later ships.

Europe, the concern now controls Alstom, Chantiers de l'Atlantique's Saint Nazaire Yard, as well as the Finnish yards and facilities in Norway and Rumania, employing a total of 16,000 workers.

Stepping back in time to the 1970s and away from Scandinavia for a while, we come across the birth of a true shipping giant: Carnival Cruise Lines. Ted Arison's brainchild pioneered the concept of competitively priced, shorter cruises that offered the onboard glitz of Las Vegas. A new slogan 'The Fun Ships' was adopted, defining the wide range of activities available that would, it was hoped, attract younger and more family orientated clientele. Arison purchased second-hand tonnage, three ships that had sailed as British-flagged liners. *Mardi Gras*, his first fleet member, had been built for the Canadian Pacific Steamship Co. as their *Empress of Canada* in 1961. *Carnivale*, five years her senior, but which he purchased in 1975, also started life with Canadian Pacific under the name *Empress of Britain*, whilst *Festivale*, bought in 1977, sailed the UK-South African route as Union-Castle's *Transvaal Castle* and later Safmarine's *SA Vaal*.

Arison's 'Fun Ship' idea was proving a real winner, thanks in no small way to his company's extremely astute marketing techniques, and he was encouraged to turn to

Royal balconies. Princess Cruises' *Royal Princess* (now *Artemis*) was the first cruise liner to feature a row of balconies. Her life tenders are shrewdly positioned on a lower deck so as not to restrict the views.

Danish shipbuilders for his first purpose-built cruise ship. *Tropicale* was delivered by Aalborg Werft in 1982 and was joined three years later by the larger *Holiday* and later two 47,262-gross-ton sisters, *Jubilee* and *Celebration*, all built at Malmo in Sweden. The Caribbean had become the cruise centre of the world and Miami the busiest cruise liner port. Yet, despite this surge in new and converted passenger vessels, NCL's *Norway*, as we have already seen, remained head and shoulders above the rest in terms of tonnage and passenger capacity. Until, that is, the momentous arrival from France of the biggest and most ambitious cruise ship yet.

Royal Caribbean's 73,192-gross-ton *Sovereign of the Seas* was the first of three sisters constructed in a building dock at Saint Nazaire. Designed to accommodate 2,682 passengers (later increased to 2,852) and measuring 268.2m in length and 32.3m width, she cost some US$183.5 million to build. If *Norway* had become a destination in herself, a concept adopted also by Carnival as they realised that making a ship a 'vacation location' rather than simply a mode of transport worked wonders for their sales, *Sovereign of the Seas* was surely the first floating resort. Two 650-seat dining rooms plus the Windjammer Café for casual dining, ten lounges, including the balconied Follies showlounge, a

Miami quickly became the world's busiest cruise ship port.

The futuristic 73,192-gross-ton *Sovereign of the Seas* brought Royal Caribbean into big-ship cruising.

well-equipped fitness and spa centre, two outdoor pools and an impressive five-deck-high Centrum (atrium) would ensure that her pampered passengers experienced a cruise holiday to remember. Christened by Rosalyn Carter, wife of US President Jimmy Carter, at Miami, *Sovereign of the Seas* was originally destined to be called *Song of the World*, but a last-minute name change was announced by video conference between Miami and Saint Nazaire in March 1986. The change proved to be a stroke of genius, for the 'of the sea's' suffix, rather like the Viking Crown Lounge, would become an instantly recognisable feature of Royal Caribbean ships over the years to come.

Monarch of the Seas and *Majesty of the Seas* were identical to *Soverein des Mers*, as she was known in her country of birth, apart from a little fine tuning. I caught up with the *Monarch* in October 1991 during her delivery voyage and was immediately awestruck by the deluge of glistening brass and glass that adorned her huge Centrum. The area was the work of Norwegian designer Njaal Eide who had created the first atrium at sea on the inspirational *Royal Princess*. Expansive atriums were becoming the vogue on all American-based cruise vessels, acting as overwhelmingly welcoming entrance lobbies. First impressions count, so they say, and there was certainly no holding back the interior designers as they were let loose on the largest of ships' public areas.

Glistening brass is in abundance within the huge Centrum on each of the *Sovereign*-class vessels.

Nearly done. Workers carry out a few finishing touches to the lido area on *Monarch of the Seas*.

With my fellow guests I climbed eleven passenger decks and continued upwards into the impressive 360-degree Viking Crown Lounge high above the upper decks. Like *Song of America*, built ten years earlier, *Monarch of the Seas* and her two sisters boasted a 250-seat observation lounge that completely encircled the funnel. The new ship was on a tight schedule to meet her Miami deadline, and although her French builders had successfully fought off competition from the Finnish yards, the sight, from on high, of shipyard workers scurrying around the decks making good the swimming pools and other activity areas gave the impression that all was not quite complete.

A pleasant luncheon in one of *Monarch*'s two main dining rooms – I cannot recall which – as the ship's hotel staff bent over backwards to impress their first guests, revealed further employment of shining brass. In the early '90s we arrived in an age of increasingly stringent fire safety regulations as SOLAS did its best to protect the fare-paying vacationer, and consequently materials of a non-combustible nature were called for. Hardly surprisingly, cruise marketing people were revelling in their descriptive phrases as they referred to such decor as 'offering a stunning effect'.

Royal Caribbean's *Monarch of the Seas* was the largest passenger ship to visit Southampton since the Cunard Queens.

I recall writing an article for a corporate publication in which I described *Monarch of the Seas'* arrival at Southampton as a moment of great significance, as she was, I was keen to extol, the largest passenger ship to visit the famous port since the legendary *Queens*, retired from service in the late 1960s. As events transpired, her call was duplicated a few months later by *Majesty of the Seas*, the third of the class, built to exactly the same specifications. Surely cruise ships could not get much bigger, I thought, a view shared by more than one eminent maritime journalist of the day, but how wrong we were all proved to be.

We were now in an era of prefabricated passenger ship construction that, coupled with computer-aided design, empowered the creation of considerably larger vessels that could be completed in much shorter times. The days of the customary slipway were over. No longer were ships under construction rising like huge skeletons from their longitudinal backbone. The change process had been gradual; as we have seen, *QE2* had been put together by using the new prefabrication method, but launched from a slipway. Now the spectacular sight of a new hull surging sternwards towards the waiting waters was a memory. Ships could be built to a more advanced state, appearing, externally, almost complete before being floated out of their building dock for final fitting out. This comparatively low-key ceremony, normally attended by shipyard workers and ship-owning and shipbuilding VIPs, would be succeeded by the glamour of the naming ceremony, normally at the ship's home port but occasionally at an interim port of call during her delivery voyage. Fear not, however, for that expensive bottle of the best, such an intricate part of the naming ceremony, has not been lost in time, and neither has the tradition of requesting a member of royalty or the wife of a national president or the shipping company's chairman to name the ship, although there is an increasing tendency nowadays to pass the honour to that unavoidable icon of the twenty-first century, the 'international celebrity'.

Externally, the classical rounded profiles of previous decades (*QE2*, *United States* and others) had been replaced by the 'wedge' shape instigated by Kloster's first cruise ships.

Perhaps the application of the term 'wedge' is being a little unkind to at least a number of modern cruise vessels whose designers have tried their best to provide an aesthetic appearance, but, as Douglas Ward observes in his excellent annual, *Ocean Cruising & Cruise Ships*, 'Designers are now required to squeeze as much as possible into the space provided.' After all, he acknowledges, 'you can squeeze more in a square box than you can in a round one.'

In years past ocean liners would be designed externally first and passenger accommodation and amenities inserted wherever internal space would allow, taking into account, of course, machinery, cargo and crew space. In the modern world, interior ship designers are brought on board from an early stage, ready to fulfil the ship-owner's ambitions and passengers' dreams without overstepping the rules and regulations of the sea and predetermined budgets. Three marine designers who were forging a fine reputation within the cruise industry in the 1990s were Robert Tillberg, Petter Yran and John McNeece. Tillberg, from Sweden, and Yran, a Danish architect, had met during their involvement with the 'Love Boats' *Island Princess* and *Pacific Princess*, the stars of the highly popular American television series of the late 1970s that cannily conveyed the romance of cruising into households across the world. They later collaborated with McNeece, who hails from Scotland, on all three Royal Caribbean's *Sovereign*-class vessels. When P&O announced their plans to order their largest ever passenger ship, the first liner to be custom-built specifically for the British cruise market, they immediately turned to the well-respected trio to design a ship that would meet their every criteria.

The seeds for the proposal were sown in 1988 and over the coming three years technical, financial and design considerations would need to be met. At the end of 1991 all the right conditions came together and P&O signed a contract with the Meyer Werft Shipyard, located in Papenburg, Germany. A public announcement of the order was made on 20 January 1992. There was an element of surprise that P&O had allocated the work to a German yard rather than going to Finland or France who were undoubtedly leading the field at the time in terms of producing sizeable cruise ships, and their new vessel would, indeed, be Meyer Werft's biggest ship to date. However, this private company, founded on the banks of the River Ems way back in 1795, was now among the most modern shipyards in the world and boasted a covered building dock that, following a recent 100m extension, measured 370m in length, was 101.5m wide and 60m high, or, in other words, ten times the size of London's Royal Albert Hall! P&O were clearly impressed with Meyer Werft's reputation as builders of many different types of ships and had the comfort of knowing that by building the ship indoors, work would continue at a level pace even through the rigours of a harsh north German winter.

With delivery planned for 1995, construction would take just two years, putting pressure on the ship's designers to ensure that drawings were correct from the very beginning. But this ship would be no ordinary cruise liner and would represent a new challenge for Tillberg, Yran and McNeece. Until then their work had involved ships destined for the American market, operating predominately in the Caribbean. P&O would be basing the new liner in Southampton, from where she would sail to various points of the compass, from northern Europe to the eastern Mediterranean and even on world cruises. She would need to be fast to reach such far-flung destinations within acceptable cruising time, and be seaworthy as well. But most important of all, she would have to appeal to the British cruising public who had yet to attune themselves to the bright decors of the American-based ships, preferring softer colourings and more intimate surroundings. For inspiration her designers needed to look no further than *Canberra*, which had been held in high esteem by P&O's clientele for many years.

Robert Tillberg, as co-ordinating architect, fixed himself a cruise on the thirty-year-old ship and was immediately captivated by her congenial onboard atmosphere. No marine architect can design something so intangible, but he (or she) can contribute to its cause by promoting the comfort and contentment of a ship's passengers. Those on *Canberra* could enjoy an easy flow around the ship, a wide range of public rooms where everyone could find a niche where they would feel at home, and a good choice of cabin types. The interior layout of Royal Caribbean's *Sovereign*-class trio is arranged vertically, with most public rooms located aft and the cabin accommodation forward. This is fine for ships principally operating on seven-day schedules, but for those offering longer voyages, even round-the-world sailings, a nice mix of accommodation throughout the ship is preferable. *Canberra* had this benefit and Tillberg and his team now knew the way forward. In October 1992 the first steel was cut and on Thursday 11 March 1993 the keel of the new cruise liner was laid.

Within the huge covered building shed Bernard Meyer, managing director of Meyer Werft, welcomed the assembled guests, observing that the projected 1995 completion of the vessel would coincide with the shipyard's 200th anniversary. P&O Cruises' chairman Tim Harris responded by announcing that 'after much thought, consideration and consultation, we have decided the right name for this ship is *Oriana*.' A popular name indeed, recalling her predecessor that served Orient Line and P&O from 1960 until 1986.

Then two shiny coins – a newly minted penny piece and a pfennig (the euro had yet to reach Germany) were placed on the first block of the ship, in keeping with the custom that is thought to have originated in Greek mythology and is intended to bring good luck to the ship, her builders and owners and all who sail on her. After a break for lunch a second section of *Oriana* was placed on the first, making the lucky coins a permanently integral part of the ship.

Over the coming months *Oriana* steadily took shape, block by block, as prefabricated sections were joined together in the building hall. Forty-five sequentially numbered blocks would make up the whole ship, from keel to funnel and mast, comprising 14,830 tons of steel. By April 1994, 87 per cent of the total steel had been put in place, and results of wind tunnel tests on *Oriana*'s funnel had confirmed its optimum shape. Three months later, on Saturday 31 July, the vessel was eased out of the covered dock that had been her home for more than sixteen months and moved to her fitting out berth. Only her funnel and mast need to be added to her external profile. Now our eminent trio of designers would see the fruition of their intricate work. All three were head of their own companies (and still are) in whose names *Oriana*'s design contracts were awarded. Robert Tillberg, whose initial brief from P&O included the deck placement of the main public rooms, was running Arkitektkontoret Robert Tillberg AB in the small Swedish town of Viken, situated just north of Malmo. Alongside him and within the same company was Anders Johansson, whose responsibility it was for the important task of monitoring the shipyard's work to Tillberg's designs.

Petter Yran's company, Petter Yran & Bjorn Storbraaton Architects AS, had been given license to design all of *Oriana*'s 914 passenger cabins. His staff of twenty was fortunate enough to work in the former Oslo boat club building from where they could enjoy stunning views across the harbour. His company could boast credits for its work on numerous cruise vessels, large and small, but almost all had been created for the North American market, so its involvement in a ship sailing out of Britain was, as Petter put it, 'very exciting'.

The first piece of the jigsaw. The keel-laying ceremony of *Oriana*.

John McNeece Ltd was responsible for designing *Oriana*'s shopping areas, the Harlequin nightclub, photo-shop, casino, Anderson's Club bar, the Pacific Lounge, a venue for evening shows and the Lord's Tavern pub, an idea taken directly from the *Canberra*. McNeece's experience of running a design business went back to 1963 when he specialised in interior design work in Scotland. Thereafter he worked on hotel facilities before entering the world of maritime design development, firstly with North Sea Ferries and then cruise ships. *Oriana* would be the first UK-based passenger vessel to receive the McNeece designer touch, in preparation for which he spent weeks touring mainland Britain to ensure he fully understood its peoples' tastes in design and decor.

The 69,153-gross-ton *Oriana* was handed over to P&O Cruises on 2 April 1995 and named by HRH Queen Elizabeth II at Southampton four days later, a fitting honour for a British ship of such significance. With a length of 260m and 32.2m wide, the ship is powered by four MAN B&W diesels, generating 47,750kw and a service speed of 24 knots, making her one of the fastest cruise liners afloat. *Oriana*'s maiden departure from Southampton on 9 April drew generous media coverage and thousands of spectators who braved a chilly, gusty wind to pack every vantage point along Southampton Water. P&O's new liner was sailing on a fourteen-day cruise to the Canary Islands, and after that into a secure and prosperous future, one that encouraged a £12 million refit in 2006 during which she was re-registered in Bermuda so that weddings at sea could be held on her. The popular Lords Tavern was further extended, a new ninety-six-seat exclusive restaurant designed by British celebrity chef Gary Rhodes was installed and she currently sails with a passenger capacity of 1,822, although there is room for a maximum of 1,928. In her initial two seasons she partnered *Canberra* until the legendary liner's shock withdrawal from service in 1997.

But with the British cruising market on continuous incline, P&O returned to Meyer Werft for their next ship, the 76,152-gross-ton *Aurora*, delivered in May 2000. A whole new fleet of European-owned cruise ships were just around the corner. But it is high time we return our attention to the opposite side of the Atlantic.

Joe Farcus is a most remarkable man. As the interior designer of Carnival Cruise Lines' ships he has single-handedly prompted the astonishing growth of the Arison empire into the world's largest cruise company. Now in his sixties, Farcus grew up on Miami Beach surrounded by tourism which, he considers, influenced his interest in the leisure industry. Even as a small child he displayed a great love for drawing, and when he was twelve he came across a little book entitled *How to Draw Merchant Ships*, a book that was to determine the rest of his life. After leaving school Farcus trained as a land-based architect but, by chance, he met Ted Arison who had just founded Carnival and found himself involved in the refurbishment of the *Mardi Gras*. Although still employed by another architect, Farcus found more time to work with the renovation of *Carnivale* and, excited by the news that Arison was buying the classic liner *SA Vaal*, he left his employer to concentrate full-time on a career in marine architecture.

Nowadays Joe Farcus is proud to reflect that 'Carnival has kept me busy since 1977'. Although today's cruise ships are more like floating towns than passenger liners, Farcus

Escorted by a flotilla of small craft, *Oriana* departs from Southampton on her maiden voyage.

still thinks it is ultimately the romance of the sea that draws people from all walks of life to take a cruise. Naturally, no one would want to return to the old days of liner travel before the advent of stabilisers and air-conditioning, but, he believes, cruise passengers do like to feel they are on a ship, and a good-looker at that.

The year 1990 was a milestone for Joe Farcus, Carnival Cruise Lines and Kvaerner Masa-Yards as the 70,367-gross-ton *Fantasy* was delivered by the Helsinki New Shipyard. The first of eight similar vessels, the largest series of cruise ships ever ordered, she confirmed a new strong alliance between Carnival and the Finnish shipbuilders that would last into the new millennium. Farcus has been allowed to stretch his skills and imagination to new limits on all the vessels, employing combinations of bold colours and creating an individual theme running through the interiors of each ship. His efforts have

Welcome to Hollywood. Humphrey Bogart and Ingrid Bergman (or rather their life-sized effigies) appropriately seated in Bogart's Café on *Fascination*.

been aided considerably by the availability of computerised lighting during recent years, enabling marine architects to develop some wonderfully atmospheric public areas. One feature that has remained consistent across all eight ships of the series is the dramatic six-deck-high atrium, topped by a large glass dome, providing a most unforgettable welcome for arriving passengers. Farcus describes his work as 'creating entertainment architecture', and the emphasis on all-round entertainment, accentuated by the glamour and glitz of Las Vegas, is really where Carnival ships score.

Fantasy and her sisters were each designed to carry a maximum of around 2,600 passengers (approximately 2,050 in lower-berths), quite a dense population for ships of their size. Of the 1,020 cabins, at least 400 are without a view, but Carnival can make the point that their passengers are likely to spend less time in their cabins than on rival cruise vessels as there is so much for them to do on board. The only occasion I rubbed shoulders with a Carnival cruise ship was my brief encounter with *Fascination*, fourth of the *Fantasy* series, during her delivery voyage in the summer of 1994. Being a typically conservative British fellow, my first impression was that of quite... no, to be honest, *vastly* overstated interior decor, but once my wife and I had indulged ourselves in a very acceptable evening meal that featured the waiters' parade of the inevitable Baked Alaska to resounding Caribbean music followed by a fast-moving show and late-night disco sounds in a neon-lit lounge, we began to comprehend why the 'Fun Ship' concept has proved such a master stroke. We returned time and again to gaze at the excellent twenty-four life-like figures of Hollywood stars of past decades that represented the theme applied by Farcus to our particular ship. Seated at the piano at Bogart's Café, for instance, was Humphrey Bogart, accompanied by Ingrid Bergman, and if we fancied a gamble then we would be greeted at the casino by Lucille Ball. Such a clever innovation, unique, as far as I am aware, to the *Fascination* herself.

If *Fascination* and her seven sisters reach the dizzy heights with their interior features, then externally they fall a little short. Although not unreasonably pleasing on the eye, the limited length of their bows ahead of the main superstructure and the squared-off stern require the provision of a swept-back funnel to offset the appearance of their ten-deck sides. Along with the smaller *Holiday*, *Jubilee* and *Celebration*, the *Fantasy*-class ships became a design template for Carnival's later, larger cruise vessels and their red, white and blue finned funnels have become as familiar in the southern United States as the sun-kissed palm trees. On the upper decks of *Fascination* I found the 'Fun' theme continuing with the provision of huge flumes linking with the ship's central swimming pool (there are three pools altogether).

Each of the Carnival octets was assembled in blocks within a covered dry dock at Kvaerner Masa-Yards' Helsinki complex. The building hall, which afforded much appreciated protection for the 1,700 workers during the freezing Finnish winter, was extended to a length of 280m in 1993, but once floated out, the ships would lie outside

Fourth of the *Fantasy* series, *Fascination* bears all the Carnival hallmarks in her external design.

Outdoor activities. The expansive lido area on *Fascination*, showing the main pool and flume.

the hall until final completion. The sisters were turned out with such speed and regularity that the sight of two white cruise ships lying alongside each other at differing stages of fitting out became as familiar for the locals as Helsinki's waterfront architecture.

All eight of the *Fantasy*-class were fitted with a diesel-electric power station system, featuring six Wartsila-Sulzer main engines, that has proved the epitome of reliability, although the final two to be built, *Elation* and *Sensation* differ from the others by being equipped with a new pod propulsion system, a highly inventive method of driving even the largest cruise liners across the oceans. The pods are rather like massive outboard motors and replace conventional propeller shafts, propellers and rudders, and even the need for stern thrusters that in themselves are recent innovations. Each pod contains an electric AC motor and weighs on average around 170 tons. The positioning of the ship's propulsion to outside the hull releases much valuable internal space for other uses. Unlike traditional screws, pods *pull* rather than push a ship through the water, thanks to their incorporated forward-facing propellers that can be turned through a full 360 degrees for unprecedented manoeuvrability (for a ship to go astern they are either rotated 180 degrees or the thrust can be reversed). This rotation procedure is referred to as 'azimuthing' within the marine industry. The pods manufactured for Finnish-produced vessels such as Carnival's *Fantasy* series are branded 'Azipods', developed as far back as

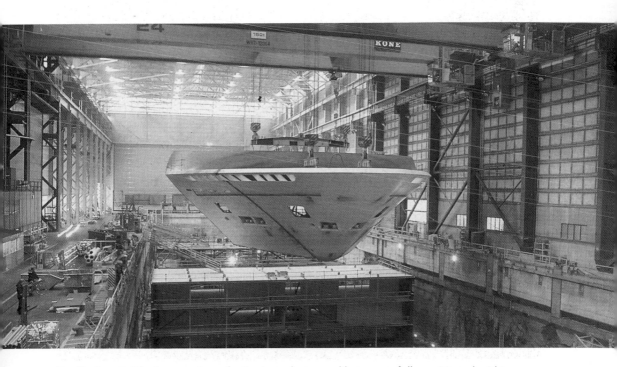

Gently does it. The bow section of a *Fantasy*-class vessel being carefully positioned within Kvaerner Masa-Yards' Helsinki building hall.

1983 by Kvaerner Masa-Yards and Helsinki-based ABB Industry and fitted to a variety of vessels, including ice-breakers. Certainly pod technology appears to be the propulsion system of the future, virtually eliminating vibration at the ship's stern and reducing its turning circle by almost one-third.

The advent of the modern cruise liner has indisputably swollen the order books of French, German and Finnish shipyards to unexpected levels, yet it is an Italian concern called Fincantieri that can rightly claim to have produced the highest percentage of modern cruise ship tonnage. As we saw in Chapter Four, Italian shipbuilding was responsible for the creation of some fine ocean liners before and after the Second World War, although Fincantieri itself is relatively young, emerging as a major passenger shipbuilder in the late 1980s. Nowadays eight facilities are operated by the group, 83 per

Twin Azipod propulsion units fitted to *Elation*. Note the forward-facing propellers that have the effect of pulling the ship through the water.

Bella Princessa. The Italian-built *Sun Princess* replaced *Norway* as the world's biggest cruise liner in 1995.

cent of which is controlled by Fintecna SpA and the remaining 17 per cent by various financial institutions. The principal facilities involved in cruise ship construction are the Montfalcone Shipyard, situated in north-eastern Italy, and the Marghera Yard, close to Venice. The Montfalcone Yard brought Fincantieri into the modern cruise era with the completion in 1990 of the 70,285-gross-ton *Crown Princess* for Princess Cruises, followed by a sister, *Regal Princess*. At the time they were among the world's largest cruise liners. Between 1992 and 1994 the same yard produced a 55,451-gross-ton trio of good looking vessels for Holland America – *Statendam*, *Maasdam* and *Ryndam* – before embarking on the construction of another new build for Princess Cruises, *Sun Princess*, which replaced *Norway* as the largest passenger ship afloat. Her entry into such a fast-growing fleet represented an impressive accomplishment by Princess Cruises who had taken great strides since being taken over by P&O in 1974. If *Crown Princess* had brought the company into the up-to-date world of floating resorts then the 77,441-gross-ton *Sun Princess* would propel it to even higher planes.

The design of the new ship would incorporate 410 balconied cabins and two separate atriums. Princess Cruises had entrusted the creation of the atriums to the experienced hands of Njaal Eide, designer of the atrium on *Royal Princess* and the 'Centrums' on Royal

He's all ears! The Mickey Mouse concept is even extended to the funnels of Disney's pair of cruise ships.

Caribbean's *Sovereign*-class ships. For *Sun Princess* he produced a four-deck main lobby and a second plaza, a two-deck atrium interlinking the casino with the discotheque on decks 7 and 8. Alongside him was Genoa-based Giacomo Mortola whose responsibility was the designing of the dining rooms, show theatre, piano bar and English pub (no, not even an English designer here!), the wine bar and pizzeria. Despite the size of the ship, both architects were successful in producing an impression of intimacy throughout many of her interior spaces. On completion *Sun Princess* was assigned to the Caribbean and Alaskan cruise circuits and flew the Italian flag. Nowadays she is registered in Great Britain and operates with a maximum passenger capacity of 2,250.

A closer study of Princess Cruises' thirty-year history further reveals the remarkable development of the company that, on delivery of *Sun Princess*, became the world's third largest cruise line. It was the charter of a 6,000-ton ferry from the Canadian Pacific Railway in 1965 that put the business well and truly on track. The ship was called *Princess Patricia*, from which the Princess company name was derived, and she was employed on cruises to the Mexican Riviera and Alaska, and with great success. The company doubled its fleet after two cruising seasons by chartering a newly completed Italian passenger ship, the *MS Italia*, renaming her *Princess Italia*. In 1968 a third ship, the SS *Carla*, was brought in on charter and renamed *Princess Carla*. In the 1970s Princess Cruises purchased a pair of 21,000-gross-ton Norwegian passenger vessels that would become the famous 'Love Boats', sealing the fame of the company (that by then was part of P&O) across the United States and beyond. Success built on success with the completion of the influential *Royal Princess* in 1984 that, as I write, sails under the P&O brand as *Artemis*. A sister to *Sun Princess*, the identical *Dawn Princess*, was delivered by Fincantieri in 1997.

I entered the floating world of Disney in July 1999. The Disney Corporation has set itself incredibly high standards over the years, especially its superbly organised theme parks and the genuine quality that oozes through everything it does. Now it was extending these attributes to the seas with the introduction of two purpose-built cruise vessels that would operate three- and four-day cruises out of Florida's Port Canaveral as part of a seven-day sea and Disney resort holiday package. But would the corporation's novel concept prove an inspirational winner or nothing better than a 'Mickey Mouse' idea? Disney certainly set off on the right foot as far as passenger ship aficionados were concerned by designing the first ocean cruise ships sporting two funnels since the halcyon days of the 1950s. Its direction was to recapture the atmosphere of the true transatlantic liner and with their funnels painted red and black and smart black hulls, the duo enjoy more than a passing resemblance to Cunard's famous *Queen Elizabeth* of 1940 or even French Line's *Liberté*, apart from the fact that the shape of Mickey Mouse's face and ears is depicted on their funnels. The forward funnel is a dummy, housing an ESPN sports bar and broadcast centre, but this is hardly new – even *Titanic* had one dummy amongst her famous four smoke stacks!

Unlike the bygone liners that Disney's handsome sisters duplicate, *Disney Magic* and *Disney Wonder* were built in Italy, further products of the industrious Fincantieri group.

It's Mickey again, this time taking the form of an upper deck swimming pool and accompanying whirlpools.

Both were actually constructed in two halves and joined together at the Marghera Yard. *Disney Magic* was completed first, entering service in July 1998, whilst the *Wonder*, which I experienced amid her delivery voyage, joined in August 1999. Each ship has 875 cabins (388 with a private balcony) that are more spacious than on most other cruise ships and, notably, twelve are designated as wheelchair friendly. The ships are veritably Disney throughout, although their respective interior decors do differ: the *Magic* is very much Art Deco, whilst the *Wonder* is better described as Art Nouveau. Highlights of both ships are undoubtedly a three-storey atrium lobby, a superb 1,240-seat 'Walt Disney Theatre' that extends through four decks and three differently themed dining rooms ('guests', as Disney likes to call the passengers, take dinner in a different one each evening). My personal favourite eating place has to be Animator's Palate where the walls of the room

The 69,153-gross-ton *Oriana* continues to cruise the world's oceans despite growing competition from cruise liners up to twice her size.

gradually change from black and white into full-colour animation as the guests dine. If a Disney tune is your requirement, then listen to the ships' whistles that play a seven-note rendition of *When You Wish Upon a Star* (I would still like to find the culprit who set off its heart-stopping tones just as I had arrived at the top of the stairway that leads down to the inventively designed Mickey Mouse-shaped swimming pool, almost dispatching me for an unexpected quick dip!).

On 22 February 2007 Disney Cruise Line, a division of the giant Walt Disney empire and owner of the two ships plus Castaway Cay, a private island cruise venue, announced it had ordered two further vessels. This time the company has turned to the Meyer Werft Shipyard at Papenburg, who built *Oriana*, to construct their biggest cruise ships yet, for the duo will be two decks taller than the existing pair, measure 330m in length and being 37m wide (as compared with the *Magic*'s and *Wonder*'s 294m x 32.2m) and have a gross tonnage of around 122,000. As testimony to Disney's customer care policy, *Disney Fantasy* and *Disney Dream* will each carry only 1,250 passengers, allowing a more than generous passenger/space ratio. More significant, however, is Disney's enrolment with the '100,000-plus club', joining the rapidly increasing list of cruise lines that have taken the 'bigger the better' principle to new levels. Cruise ships of more than 100,000 gross tons are fast becoming the norm rather than the exception and, what is more, these megaships appear to be getting bigger by the day.

CHAPTER SIX

MEGASHIPS OF THE SEAS

The term 'mega', according to all the best dictionaries, means either 'large' or 'a million'. It seems to have come into vogue over the past decade, not only in everyday speech (in the UK, at least) but within maritime circles, especially since the advent of the gigantic cruise ships that sail our seas today. In this connection it most certainly refers to size rather than quantity; a 1 million-ton passenger ship is surely beyond our current comprehension and would be impracticable as well, for what present shipyard or port of call could possibly handle such a monster? Admittedly, new berths and even small private islands have been specially created to accept the latest breed of floating cities, so who can guess what the future holds? As far as this chapter is concerned, however, I apply the 'mega' term to cruise vessels of over 100,000 gross tons.

During a recent port visit I overheard a cruise industry employee of substantial knowledge referring to a certain 70,000-ton passenger ship as a 'workhorse of the Caribbean.' His comment was doubtless intended to be complimentary rather than disparaging but does demonstrate how our attitudes have changed as cruise ship growth has remorselessly moved on. While ships of that size plough their honest furrows on the Caribbean and beyond, vessels two or even three times larger are now making all the nautical headlines as they embark on the cruising scene.

The demise of the full-time transatlantic liner finally put an end to the fierce competition that had persisted between the principal shipping nations, and also premier ship-owning companies, in their bids to own the world's biggest passenger liner. In the twenty-first century the majority of the largest vessels are, in fact, operated by international conglomerates and, although the owning concern of the very biggest must reap undeniable pleasure in holding that accolade, the decision to create such a ship is mainly determined by modern-day economic factors.

By complying with the maxim 'economy of scale', cruise ship operators are able to reduce their overheads per passenger, and by taking this to its very limits are finding themselves in charge of passenger vessels of unprecedented dimensions. Furthermore, as the expectations of cruise clientele grow, ship designers are under increasing pressure to meet passengers' wildest fantasies with amenities such as huge auditoriums and ice rinks that require an ocean-going structure the size of a city to encompass. A disadvantage of their immensity is their width, being too wide to be Panamax (capable of transiting the Panama Canal, which is 33.53m wide), thus somewhat limiting their routes.

It was surely fitting that the first 100,000-gross-ton passenger vessel should have been built for Carnival Cruise Lines, as they were, and still are today, the world's largest cruise organisation. The 101,353-gross-ton *Carnival Destiny* was delivered in 1996 by the Fincatieri Shipyard at Monfalcone, having taken, quite amazingly, just twenty months to complete, an achievement accomplished by the employment of the latest building method of prefabrication and three-dimensional computerised design. More than 500 separate suppliers were involved in the ship's creation that cost, in all, some US$400 million.

Within the brochure issued in celebration of the new ship, Carnival's statement that 'we didn't plan to make her so big, it just came,' bears out the philosophy just analysed. Twelve passenger decks provide 30,000cu.m of public interior space, permitting interior designer Joe Farcus to produce passenger cabins more spacious than on most other cruise vessels, together with some wonderfully expansive public areas. He was prepared to admit at the time that not everyone enjoys being on a ship of such magnitude, but he was confident that such a rich combination of his own vast experience and modern-day techniques would soon encourage these people to change their minds. 'On the *Destiny* we have made the best use of the great availability of space,' he stated.

Designed to accommodate a maximum of 3,340 passengers in 1,320 cabins (2,640 'guests', if only the lower beds are occupied) of which 418 each have a private balcony, *Carnival Destiny* is, without argument, a ship of superlatives. Her passengers have no trouble in losing themselves in the dream world of her three-storey Palladium showlounge, two twin-storey dining rooms seating a total of more than 1,700 people at one sitting, four swimming pools and seven whirlpools, not to mention a 60m outdoor slide, even longer than the flumes found on Carnival's *Fantasy*-class ships. Farcus has wisely, in my opinion, added more subtlety to *Destiny*'s interior décor, and she is noticeably less glitzy than earlier ships of the fleet, consequently, I would think, appealing to a wider audience. The ship received a multi-million-dollar refit in 2005 to keep her in line with her more recent competitors.

Carnival Destiny was, on completion, the first passenger ship to exceed the gross tonnage of the 1940-built *Queen Elizabeth*. Further comparisons between the two vessels serve to accentuate their vastly contrasting external profiles. *Queen Elizabeth* was just over 315m long and 35.96m wide. She had a long foredeck and her sleek superstructure descended in gradual steps towards her stern. *Carnival Destiny*, on the other hand, measures 272m in length and is 35.3m wide. Her high superstructure extends forward almost to her foreshortened bows and slants sharply down to her stern. Gross tonnage is a measurement of cubic internal space, not weight, and consequently the more spacious *Destiny* boasts a much higher gross tonnage than the older liner.

Happily, Carnival's co-founder, Ted Arison, lived to witness his company's first 100,000-ton ship, but passed away in October 1999. One of his legacies was the forming of top basketball team Miami Heat in 1988, and, following his death, son Micky became its major shareholder. Today Micky Arison holds the posts of chairman and CEO of the Carnival Corporation and is listed as one of the world's richest businessmen.

As *Carnival Destiny* sailed out of Fincatieri's Montfalcone Yard, an even larger cruise liner was taking shape behind her. Princess Cruises, P&O's American subsidy, were paying the Italian builders $450 million to construct the first of three *Grand*-class ships whose thirteen passenger decks and extra length and width (290m x 36m) would award her a gross tonnage of 108,808, ousting the *Destiny* as the world's biggest passenger ship. Of her 1,300 cabins, no fewer than 710 were designed to incorporate private balconies, rows of which dominate her external appearance, that, with her sheer transom stern, produce a rather 'boxy' profile that is somewhat offset by a low, streamlined funnel.

The private balcony (or veranda), first introduced on Princess Cruises' 1984-build *Royal Princess*, has proved to be one of the most successful and innovative features of the modern-day cruise ship. Anyone who has sampled this added pleasure would settle for nothing less, although it should be highlighted that a number of vessels currently sailing our oceans fail to provide partitions that totally reach from floor to ceiling between their balconies, somewhat spoiling their occupants' privacy.

Grand Princess entered service in May 1998, to be later joined by two identical sisters, *Golden Princess* (May 2001) and *Star Princess* (February 2002). All three have proved to be successful additions to the cruise circuit, *Grand Princess* leaving her Caribbean base for Europe on occasions. Indeed, in summer 2009 she operated a series of Mediterranean cruises out of Southampton. During their construction the trio's balconies were built outside the main body of the ship rather than within the deck area, thus allowing increased interior space. An amenity unique to the class is a glass-walled Skywalkers nightclub suspended almost 46m above the stern. The positioning of this feature has produced an overall external profile that has failed to meet with universal taste. Some critics have unkindly referred to its appearance as a 'shopping trolley'. Personally, I would rather regard it as a spoiler on a sports car. From on board, however, the reason for the unusual design becomes more apparent. As I discovered on *Golden Princess*, a travelator transports passengers from an upper deck to the nightclub that spans the whole width of the ship and doubles as a superb observation lounge by day. The downward view from the adjoining deck to waterline level is not for the vertigo sufferer, though!

Each of the three vessels is equipped with four swimming pools, one of which is covered by a magradome. They are also blessed with a wedding chapel where, thanks to the ships' Bermudan registry, American couples can legally tie the knot. During my time on *Golden Princess* I was pleasantly surprised to find that, in addition to an inevitably huge show lounge and three expansive restaurants, a handful of cosy alternative eating places and quiet corners belied the magnitude of the rest of the ship.

Princess Cruises' first megaship, *Grand Princess*, sailing on her inaugural cruise in 1998. Unfortunately her 'rear spoiler' design has been the butt of criticism over the years.

Looking along the travelator that leads to the nightclub located over the stern of *Golden Princess*.

As *Grand Princess* was fitting out in Italy, Royal Caribbean International was signing a contract with Finnish shipbuilders Kvaerner Masa-Yards for the construction of a cruise ship of even greater proportions. *Voyager of the Seas* would be the first product of a new *Eagle*-class concept of which two vessels were originally planned, with an option for a third. In point of fact, five such ships were built in as many years as demand for cruise holidays continued to spiral. Up until then the size of megaships had edged just over the 100,000 gross ton mark, but *Voyager of the Seas* was taking cruise liner magnitude to another level with a predicted gross tonnage of 142,000. In reality her tonnage fell just short of that figure, but, with her own shopping and entertainment mall, ice rink and rock climbing wall, she was clearly leading the way to the creation of true floating cities.

In a promotional pamphlet Royal Caribbean announced: 'Imagine everything you might be able to do on the world's biggest cruise ship, and then keep on imagining. *Voyager* is 30 per cent bigger than any other cruise ship because that's how much space we needed to fit all the innovations on board.' Twenty years earlier the converted *France* was being hailed as the first 'destination at sea', but still retained much of her ocean liner ambience, possibly because of a lack of variety in respect of onboard amenities. Royal Caribbean's new giant was designed around a multitude of amenities that could take the passenger the entire cruise to experience.

Building work began on Yard No.1344 in September 1997 and she was floated out from her vast building dock at the Turku shipyard on 27 November 1998, just fifty-seven weeks later. Following a series of brief promotional cruises, *Voyager of the Seas* entered full commercial service on 21 November 1999. However, her constructional journey at the shipyard had not been smooth, for she suffered a fire that damaged eighty cabins and the main restaurant, necessitating their replacement.

Voyager of the Seas has a gross tonnage of 137,280. With a length of 311.1m (1,020ft) she was, on completion, the first purpose-built cruise ship to exceed 1,000ft, yet she still could not match the 1,035ft length of the *Norway* (ex-*France*). But consider their respective beam measurements and comparisons become even more dramatic: the *Voyager* is 47.4m (155½ft) wide compared with *Norway*'s mere 34m. Even more remarkable are her high-tech specifications. Even the most stubborn passenger ship traditionalist cannot fail to be impressed with the following statistics: *Voyager of the Seas* is equipped with a huge diesel-electric power station, incorporating six Wartsila main engines and producing 75,600kw. Her propulsion machinery consists of three pod units, the outer two of which can rotate a full 360 degrees, whilst the centre unit is fixed. At her bows the *Voyager* is fitted with four thrusters, enabling her to manoeuvre independently within port and to move sideways at a speed of 3 knots. Moreover, an important design feature provides the ship with a high level of power plant redundancy. The machinery plant has been divided into two separate parts, known as the 'half-ship' concept, ensuring that *Voyager* retains at least half her operational machinery at a time of breakdown. Even telephone, public address and alarm systems have their own in-built redundancy arrangements.

As we move into the *Voyager*'s passenger areas we come across its true centrepiece, a four-deck-high boulevard from which spin off the vessel's shops and numerous entertainments.

The Royal Promenade, a focal point to which passengers find themselves drawn time and again (and proves a real godsend when trying to navigate around the huge ship), runs a length of 120m and is flanked at either end by a pair of eleven deck-high 'centrums.' Supposedly based upon London's swish Burlington Arcade, it is, for me, one of the best design features on the ship, for its architects have adroitly avoided the creation of straight lines, designing a winding effect that suppresses the impression of the mall's considerable length. This effect is further enhanced by the provision of bay windows to the three decks of cabins that give internal views down on to the Promenade. The whole idea was unashamedly (but extremely successfully) borrowed from a duo of 58,000-ton Baltic ferries, *Silja Serenade* and *Silja Symphony*, that, on entering service a decade earlier, were the world's largest of their type and were built at the same Turku shipyard as *Voyager of the Seas*.

The provision of a genuine indoor ice rink and a rock climbing wall confirms Royal Caribbean's quest to attract a wider age-range of cruise clientele. 900 seats surround the rink that allows the uniquely contrasting experience of ice skating in the hot Caribbean (rather like those ski slopes that now exist in Dubai) and adds another dimension to the ship's array of evening entertainment – the ice show. *Voyager of the Seas* was the first cruise liner to boast this amenity, as well as the rock climbing wall which is affixed to the aftermost side of the ship's funnel and, with its popularity reaching new heights, is now featured on every Royal Caribbean cruise ship.

In the year 2000 *Explorer of the Seas*, a product of the original two-ship order with Kvaerner Masa-Yards, was delivered, overtaking the *Voyager* by all of 28 gross tons as the world's biggest cruise liner. The 'optional' third *Eagle*-class ship was delivered in 2001 as the *Adventure of the Seas*, followed by *Navigator of the Seas* (that has spent recent summers in UK and European waters) and *Mariner of the Seas* in 2003 and 2004 respectively. With the help of interior designers Robert Tillberg and Petter Yran, together with Howard Snoweiss (USA), Njal R. Eide (Norway) and Thomas Tillberg from Sweden, Royal Caribbean was cementing its position as the second largest cruise line thanks to this quintet of megaships. In addition, three 90,090-gross-ton *Radiance*-class vessels were delivered by Meyer Werft's Papenburg Yard between 2001 and 2004. Royal Caribbean was expanding its fleet at the rate of knots, but was still unable to close the gap on Micky Arison's Carnival Corporation, which seemed to be growing relentlessly by the day.

The corporation already controlled Carnival Cruise Lines, which had added a pair of sister ships to *Carnival Destiny*, but a sequence of mergers and takeovers resulted in the giant concern operating more than sixty ships and thirteen different brands by 2005, including Cunard Line, acquired as a one-ship company (the *QE2*) in October 1999. The most significant merger was undoubtedly that of Carnival's amalgamation with P&O Princess Cruises in 2003, also bringing the German businesses A'Rosa Cruises and Aida Cruises, as well as P&O Cruises, P&O Cruises Australia, Princess Cruises, Ocean Village and Swan Hellenic on board, joining forces with Costa Cruises (of Italy), Holland America Line, Seabourn Cruise Line and Windstar Cruises, which were already under Carnival's expansive wing. Thankfully these cruise lines have retained their independent identities, including their funnel and hull colours.

Now, what could be the connection between sticky-back plastic and the *Queen Mary 2*? There is no need to worry, the giant cruise liner is held together with something much stronger than that, but there is indeed a link that goes back as far as 1967. A certain young lad was watching an edition of the long-running children's television programme *Blue Peter* that was being screened from the *Queen Elizabeth* in the English Channel. The youngster was immediately smitten by the sight of the famous old liner and two years later was delighted to find himself on board the brand new *QE2* after his parents had arranged a special port visit whilst the ship was in Southampton. But his interest was even further fuelled in 1972 when the same programme covered the demise of the old *Queen Elizabeth* that, as *Seawise University*, had caught fire in Hong Kong Harbour, commenting that a comparable ship would surely never again be built. But Stephen Payne, now twelve years old, thought otherwise and received a *Blue Peter* Badge for writing to the programme's production team to tell them so. He was determined to become a naval architect and, after working his way through the ranks, joined the Carnival Corporation with whom he contributed to the design of their *Fantasy*-class ships and was appointed Project Manager for the construction of Holland America's sixth *Rotterdam* of 1997.

Stephen Payne was an avid admirer of the previous *Rotterdam*, built in 1959 and held in classic-liner status with her revolutionary engines-aft design and narrow twin uptakes.

Designed by Stephen Payne. The twin-funnels of *Rotterdam* (1997) were inspired by her 1959 predecessor.

He was able to replicate, to a degree, this funnel design on the new ship, which, as I can personally confirm, is a cruise vessel of high quality. On their acquisition of the Cunard Line, Carnival Corporation decided that the lonely *QE2* should have a transatlantic running-mate that would also augment her worldwide cruising schedules. But, in line with that popular maxim, economy of scale, she would be bigger – twice as big, in fact. His appointment as chief designer for Project Queen Mary was a dream come true for Payne, and following the signing of a $780 million contract with Alstom Chantiers de l'Atlantique on 6 November 2000 (the shipyard had merged with Alstom in 1976) he set up a site-based team of project workers at the Saint Nazaire Yard whilst assuming overall responsibility for the external design of the new liner. In order to arrive at the perfect design, Stephen Payne deliberately took and adapted design elements from several famous predecessors: the bridge front from the *Queen Mary* of 1936; the Promenade Deck – open at the sides and rear but enclosed at the forward end – from the 1959 *Rotterdam*; the inclusion of a central passageway, linking the main public rooms, from *Normandie*, and, naturally; a number of ideas from *QE2*, in particular her funnel and mast profile, bow shape and the enclosed bridge. He was surely on to a winner.

Wednesday 16 January 2002 was a time for celebration as the first steel for the new liner was cut. In a speech at the ceremony, Pamela C. Conover, president and chief operating officer of Cunard Line, recalled the days of French and British Atlantic rivalry; how *Queen Mary* and *Normandie* vied for the Blue Riband. The shipbuilding company's chairman and CEO Patrick Boissier replied by acknowledging that:

> Queen Mary 2 is the occasion to go 'back to the future'. This vessel, whilst it is unique in its design, renews the prestigious traditions of the great transatlantic liners, nearly a hundred of which were built here in Saint Nazaire. Some of whom have entered the annals of history, with names like *Ile de France*, *Normandie* and *France*. But, with the completion of the latter, the tradition was broken for some forty years. Today we return with the *Queen Mary 2*, but we also integrate the best of today's technological innovations, developed here for the markets of sophisticated vessels such as cruise liners.

Nostalgic pleasantries indeed that were exchanged again later at the keel-laying ceremony on 4 July 2002, 162 years to the day from the maiden departure from Liverpool of Cunard's very first ship, the *Britannia*. As Captain Ronald Warwick, *Queen Mary 2*'s designated first master, gave the command, a 600-ton section of the liner's keel was laid on the blocks within the massive building dock, the first of ninety-eight prefabricated sections that would make up the whole ship. It was a far cry from the early days of prefabrication when a block of 50 tons in weight was deemed to be huge.

The pleasantries were over and now the pressure would be on Alstom Chantiers d'Atlantique to meet Micky Arison's testing demand that the ship would be completed in two years. The cost of running over schedule for just a single day would be £300,000. Yet without careful and precise planning, design and numerous simulated tests that all had to be correct down to the smallest detail, carrying out the constructional work at the required rate would be impossible. Consequently much pressure rested on the shoulders of Stephen Payne and his project team together with the shipyard's Senior Designer Jean-Jacques Gatepaille. Although numerous ideas had been taken from several of the new liner's illustrious predecessors, there were other considerations to be overcome.

Queen Mary 2 was going to be the biggest transatlantic liner ever built as well as an all-round cruise ship and, as we have already observed, the ferocity of the North Atlantic has to be combated by extra hull strengthening and design features if a vessel is to survive many seasons on that demanding route. Although the new Cunarder would be avoiding the wild winters of northern seas by cruising in warmer climes, the risk of the rare but deadly rogue Atlantic wave could in no way be ignored. Consequently she would have the thickest hull plating ever used in passenger ship construction, more than 2.6cm thick, and feature a breakwater barrier on her foredeck. With regard to her dimensions, her height would be limited by her need to pass under New York's Verazzano Narrows Bridge. This in turn produced a problem with the design of her funnel, and specially-contrived wind scoops were added to ensure that emissions were directed well clear of her decks. Queen Mary 2's length, moreover, could not exceed that of the ship's turning area at the port of Southampton and, in accordance with pre-arranged cruise schedules, it was initially decided that she should be no wider than the Panama Canal.

Original construction plans called for the incorporation of aluminium in the liner's superstructure as a weight-saver, but growing proof that aluminium suffers metal fatigue after some thirty years encouraged a change of mind, and steel was used in the construction of the whole of her superstructure. But the additional upper weight that resulted would have to be counteracted by a wider beam in order to overcome top-heaviness, and consequently Cunard (and Carnival) found itself with an even larger ship than planned (rather as Carnival Cruise Lines had with their Carnival Destiny) that, at 41m wide, would be unable to transit the Panama Canal after all.

Queen Mary 2 would be powered by four huge Wartsila 16V46 environmentally friendly diesel engines of such size that the liner's hull had to be constructed around them. Yet, during an intensive tank test carried out in the Netherlands using a scale model of the ship in simulated sea conditions, it was found that the diesel-electric plant alone would drive the cruise liner at little more than 24 knots rather than at almost 30 knots, the top speed needed to maintain the six-day transatlantic schedule then operated solely by QE2. More energy had to be generated and the problem was solved by adding a pair of gas turbines, small and lightweight compared with the remainder of the ship's machinery but requiring copious amounts of air to function. But where would they be positioned, for if they were placed in the bowels of the ship, internal air shafts that would take up valuable internal space would have to be installed? Stephen Payne and his team had the answer: they could be located behind the liner's funnel where they would have full access to outside air. They are now an intricate part of Queen Mary 2's external profile.

With full preparations now firmly in place, the Alstom Chantiers d'Atlantique Shipyard could proceed très vite with construction work, and on 21 March 2003 Queen Mary 2 was floated out of her building dock, but bereft of her four Merland pods (two turnable, two fixed) that had failed to arrive on time. Unfortunately the pods, after fitting, would continue to cause trouble, but ultimately they would not disappoint, ensuring smooth, vibration-free service that, in conjunction with the ship's three bow thrusters, would allow *QM2* to make a complete turn within her 345m (1,132ft) length.

The pods were eventually delivered and *Queen Mary 2* was towed out of her building dock on 2 May 2003 for fitting out in Basin C. At least her two pairs of stabilisers had arrived in time, huge affairs, each with a surface area of some 15sq.m and extendable to 6m from the ship's side. It was claimed they would reduce roll by 90 per cent, making her more stable than any previously built liner. Externally she appeared virtually complete, the lower portion of her funnel, without the 1m black top section, having been fitted on 13 March and the signal mast put in place the following day.

Three thousand fitters were involved in the ship's fitting out that included the installation of 1,500 miles of electric cable, 80,000 lighting points and the application of 200,000 litres of black and red paint on her hull. *QM2* was steadily turning into an elegant queen, just as everyone had hoped, and at 17.30 hours on 25 September 2003 she was manoeuvred from Basin C by five tugs for her first taste of the sea. During her trials the 148,528-gross-ton cruise liner handled superbly but experienced one unexpected problem as the door of one of her bow thrusters suddenly blew off. It was a problem that could only be addressed by dry-docking her, and she returned to the shipyard for repair. While she lay in dry dock, Alstom Chantiers de l'Atlantique officials had the bright idea of arranging an open day for shipyard workers and their families as a special *merci* for completing their largest-ever ship in good time. But tragedy struck on that day of 15 November when a gangway, crowded with forty-eight excited people, collapsed, pulling down supporting scaffolding and sending the people 15m down to the pavement below. It was confirmed that fifteen people had lost their lives and at least thirty were injured, and the shipyard was subsequently closed for a day of mourning.

Cunard announced that *QM2*'s maiden voyage would take place as scheduled and the liner departed for Southampton amid a vibrant atmosphere. On a wet and blustery Boxing Day morning, thousands of sightseers gathered at every possible vantage point along the Solent and Southampton Water to witness the arrival of the largest liner ever to fly the Red Ensign and, at the time, the biggest passenger ship in the world. Following a number of promotional events, *Queen Mary 2* was officially named by HM Queen Elizabeth II at a grand ceremony held on the dockside adjacent to the QEII Terminal. The Queen narrated the words of the traditional blessing, just as her grandmother had at the launching of the first *Queen Mary* some seventy years earlier. A champagne bottle smashed against the ship's bows and the ceremony closed with an impressive firework display. On 12 January 2004 the floodlit *Queen Mary 2* received a huge send off as she departed in darkness on her maiden voyage, a sell-out cruise to Fort Lauderdale. Since that day respect for the Atlantic's newest monarch has never waned. Her lucky coins are looking after her well.

The *QM2* has retained her place as the largest passenger ship to operate 'line' voyages to this day, but in the same month as her momentous maiden voyage, metalworkers

The magnificent *Queen Mary 2*, the largest ship ever built for transatlantic 'line' voyages.

in Finland were busy preparing steel for a cruise ship that would, for the first time in history, exceed 150,000 gross tons. The Turku shipyard of Aker Yards (formerly Kvaerner Masa-Yards) had been contracted to build a follow-up to Royal Caribbean's successful *Voyager*-class vessels that would, at 338.91m (1,111.9ft) be too long for their building hall and so would be assembled outside. *Freedom of the Seas* would be the first of a new three-ship *Freedom* series (originally known as *Ultra Voyager*-class) and would carry 4,370 passengers and 1,360 crew. With a total of eighteen decks, she would tower to a height of almost 64m and be the equivalent of three and a half football pitches in length.

According to the shipyard's Project Manager, Toivo Ilvonen, the huge ship would take no more than two and a half years to build, the work overseen by the vessel's first captain, William (Bill) S. Wright, an American who had lived in Oslo for almost twenty-five years. He watched ship workers join together the 172 steel sections that made up the ship, all having been created indoors away from the winter cold, and by July 2005 sufficient numbers of blocks were assembled for the ship to be recognisable. The cutting of 28,000 tons of high-quality steel plate was computer managed, but all upper decks were built of strong, light aluminium, meaning that the ship's stability was centred on Deck No.4, her strength deck.

An independent facility that operates alongside Aker's Finnish shipbuilding yards is a specialist manufacturer of modular ships' cabins, as well as hotel rooms, bathrooms and shower units. Located at Piikio, about an hour away from the Turku yard, it supplied cabins not only for the *Voyager* ships but earlier Royal Caribbean vessels and Carnival's *Fantasy* series, constructed at Helsinki, and *Freedom of the Seas* would be no exception. Using a simple but highly efficient procedure, the modules arrive at the building dock completely ready to be plumbed and wired in, an exercise that remarkably takes less than an hour for each cabin.

With her machinery installed, *Freedom of the Seas* was ready for floating out of her building dock by late August 2005, and such was the speed of her fitting out process, she was ready for her first sea trials by the following December. By then Captain Wright and his senior crew had spent valuable time at the Star Center in Florida where they were able to train in the handling of their new megaship in simulated conditions. There they could make all their mistakes and be faced with situations far more testing than in real life. As 2005 came to a close *Freedom of the Seas* left ice-bound Turku on her initial trials. Led by two ice-breakers and two tugs, she sailed into the Baltic Sea where a problem with one of her three Azipods came to light. She would have to dry dock again for the pod to be repaired, but this was already occupied by the Royal Caribbean's next *Freedom*-class ship, *Liberty of the Seas*, in early days of construction. Only at Hamburg was there was a dry dock big enough to take the new vessel, but that was four days away so it was decided that the voyage to the German port should be used as a further sea trial.

Following the repair and other finishing touches at Hamburg, the 154,407-gross-ton *Freedom of the Seas* was handed over to Royal Caribbean International and Bill Wright was delighted to take her to his home city of Oslo for promotional festivities. From there she called at Southampton and was officially named at New York on 12 May 2006. From May 2009 she has been deployed on cruises out of Port Canaveral. *Liberty of the Seas* followed her into service in May 2007, and the third of the series, *Independence of the Seas*, made her bow in April 2008 at Southampton from where she operated full seasons of cruises during the summers of 2008 and 2009. Such was their success, she is returning to the UK in 2010. All three are virtually identical, boasting the *Voyager*-class amenities of an indoor ice-skating rink, rock climbing wall and a central promenade of shops and places of entertainment, with the addition of an interactive water park featuring a Flow Rider onboard wave generator for surfing. The gross tonnage of *Independence* is actually 154,497, meaning that until late 2009 she was technically the world's largest cruise liner.

On 11 December 2007 a 60-ton, 47m-wide hull section was carefully lowered into Aker Yards' Turku building dock. The ceremony signalled the beginning of the construction of two Royal Caribbean cruise ships that would each measure almost 70,000 gross tons *more* than a *Freedom*-class megaship. Dubbed the Genesis Project, plans for the 362m-long (1,187ft) giants involved some 500 designers, 3,000 basic design drawings and 30,000 detailed engineering drawings from the supplier network. An important factor concerning the duo is that they both represent a new concept in cruise ship design, rather than being enlarged versions of previous Royal Caribbean vessels. They are both designed to be energy efficient and to create minimal waste so as to make them as environmentally compatible as possible. Additionally, the ships' passenger amenities are divided into seven individually themed 'neighbourhoods' – this being the first time the idea has been employed on any ship, confirming their status as veritable floating cities.

Cabin accommodation is predominately located within two superstructure 'towers', extending lengthways along each side of the ship and linking with the forward superstructure where further cabins as well as public areas are situated, although at the aft end, the location of a freshwater 'Aquatheater' pool, the two 'towers' remain separated. Running amidships between these cabin blocks is an open air Boardwalk extending from the stern pool towards the centre of the ship on Deck 6, and Central Park, inspired by the Manhattan landmark, that consists of landscaped gardens and a Rising Tide Bar that moves up and down through three decks, and several restaurants. Lower down, on Deck 5, is the Royal Promenade, a continuation of the idea featured on *Voyager* and *Freedom*-class ships, but, of course, on a grander scale.

The two megaships have 2,700 cabins, allowing a normal cruising capacity of 5,402 passengers, although this can be increased, with three or four to a room, to a maximum of 6,300. The advantage of the two cabin 'towers' is that the majority of cabins look either seaward or over one of the neighbourhoods. Indeed, 74 per cent of all cabins have a balcony and there are no less than thirty-seven categories of accommodation, from high-level two-storey Loft Suites downwards. There are 324 cabins (254 with balconies) overlooking Central Park, which itself rises through five decks. Even the vessels' life tenders are carried on a low deck to allow unobstructed sea views from the outward facing balconied cabins.

Not surprisingly, much hype surrounded the announcement of the Genesis Project, and although the popular opinion indicated that the first ship would be called *Genesis of the Seas*, a United States newspaper ran a 'Name That Ship' competition for the best suggestions for the two ships' names. Out of more than 91,000 entries a fifty-three-year-old Detroit resident won, proposing the names *Oasis of the Seas* and *Allure of the Seas*. So that December day was witnessing the positioning of the first of 181 sections that would make up the *Oasis of the Seas*. It was reported that the largest block would measure 22m in length, be 30m wide and weigh 600 tons. The cutting of the steel plating is a fascinating process: all the cutting is carried out under water, and once computer software has calculated the point where it should be cut, the steel is struck by a plasma-cutter, creating a temperature of 10,000 degrees centigrade at the point of the plane.

Great accuracy must be exercised in the positioning of all the sections, especially the keel. Although nowadays the ceremony of laying a passenger ship's keel does not necessarily involve the placing of the most central bottom section, even a millimetre discrepancy could affect the stability of the ship, and nowadays the use of laser beams confirms when the keel is in perfect position. Further sections are then welded to the keel and to each other as the ship begins to take shape. As another huge section is swung by crane into the building dock, ship workers direct proceedings until the block nestles in place. Then the computer takes over from the human eye, ensuring the new block is precisely in position before being welded to its adjacent sections. Once again there can be no room for error, for even a single faulty weld in many thousands of square metres of ship's structure can be a recipe for disaster once the vessel is at sea. Modern shipbuilding techniques are impressive, nobody can deny that, but there is still no accounting for the bitter cold of Finnish winters. Yet, through those snow-covered days the ship workers simply carried on – only if the temperature dropped below -20 degrees centigrade did work stop, and that was because the equipment ceased to function!

On 22 November 2008, millions of litres of water were released into the Turku shipyard's building dock and *Oasis of the Seas* was floated out and moved to her fitting out berth. With the dock now vacant, the keel of *Allure of the Seas* was laid on 2 December, only a matter of days later, and the building process began once again. Fitting out of *Oasis* proceeded on schedule throughout the winter and into the summer of 2009 when she was ready for trials. Her home port of Everglades in Florida's Fort Lauderdale readied itself for the grand new arrival and arrangements for an introductory four-night 'Labadee Extravangaza' cruise on 1 December to Royal Caribbean's private destination at Labadee on the island of Haiti. *Oasis of the Seas* was then due to begin her inaugural season with nineteen consecutive seven-night Eastern Caribbean cruises until May 2010 when she would alternate this programme with a seven-day Western Caribbean itinerary that would feature calls at Jamaica's historical port of Falmouth. In partnership with the Port Authority of Jamaica, Royal Caribbean International has created a new cruise pier and associated infrastructure at Falmouth capable of handling the two *Oasis* megaships and their thousands of passengers. A new era of gigantic cruising cities has well and truly begun.

CHAPTER SEVEN

SHIP SHAPES AND INTERIOR FASHION

We have journeyed together through almost two centuries since the little *Savannah* sailed into Liverpool from Nova Scotia, unwittingly opening the door to decades of relentless competition between the world's principal shipping nations to *build the biggest*. If a time traveller could transport himself back almost as far and stand alongside Isambard Brunel, he would witness dramatically important evolutionary changes: ships powered by engines after generations of sail and hulls of iron, and later steel, following many centuries of wood. Without these changes we could not have advanced to where we are today. Propel our time traveller a similar period forward and the thought of what he will be seeing is positively mind-boggling, even quite frightening. Automated islands could be moving continuously around the world, like huge resident ships, as highly populated countries become too crowded for comfort. The size of ships will surely continue to be a major issue, but for how long?

As the first decade of what still seems a new millennium comes to a close, and my cynicism increases with age, I have to query whether we are all pulling together in the right direction. We are in the midst of a serious global recession, one that appeared to creep up on us with little warning, and one that I pray will not affect us for as long as this book remains on the shelves. Shipbuilders' books are still healthy with cruise ship orders, but these had been placed before the economic downturn. Royal Caribbean International is committed to pay almost $2.6 billion for its latest megaships, the *Oasis*-class duo that were planned and designed when the international cruise market was growing at 10 per cent per year. In 2009, the delivery year of the first of these, the company expected a fall in net yield of much more than that. What an unpredictable world we live in!

But we must be positive. Life is cyclical, as I have found over my sixty or so years, and two world wars and the 1930s' global recession have all passed since Germany and then Britain created their famous four-stackers, the first 'superliners' capable of transporting thousands of emigrants, as well as the aristocratic rich, across the turbulent waters of the North Atlantic. Their black riveted hulls represented power and strength, a straight stem, curving well below the waterline where it meets up with the double bottom. Gone was the clipper bow, represented on Brunel's *Great Western* and sail-powered ships before her, saving much effort in craftsmanship and giving the mighty liners a more formidable look, even in times of war. A relatively straight stem prevailed even after the four-stacker had fallen out of fashion. A slight angle was awarded to that of the *Queen Mary*, whilst a few more degrees were subtly added to *Queen Elizabeth*'s stem, as three funnels gave way to two. Yet more angle was applied to the racier *United States*, but all these ships, with their long foredeck, raised bridge ahead of the main superstructure and aesthetically slanted funnels were pleasurable to the eye without the need for the extra curvature of a clipper bow, now an intricate part of modern cruise liner design.

As marine architect James Gardner observed at the time of *QE2*'s launch in 1967:

> The grace of the *Mauretania* and her sister ships was a natural result of the shipbuilding practices of their time. A sweeping sheer line, low superstructure, characteristic vent cowls and tall smoke stacks added together produced, with a little attention to symmetry and tidiness, a beautiful ship, without gimmicks.
>
> The *Queen Mary* which followed was becoming more hotel and less ship. Some care was put into her deck houses, window layout and bridge front. Giant smokestacks, still the symbol of status and power, added to her dignity. What was lost in grace was gained in scale. She was the top ship for the top people and she gained the Blue Riband. Britain was proud of her.

The reasons for the popular inclusion of the clipper bow on the latest breed of cruise vessels are twofold. The bulbous bow, that ingenious invention that came to the fore on the 1930s German liners *Bremen* and *Europa*, reduces drag ahead of a ship, thus improving speed as well as stability. This feature was enlarged by some 12m on *Queen Mary 2* once tank tests showed that a higher service speed would consequently be attained. A bulbous bow, though, would extend beyond a stem of little curvature, hindering forward navigation, so the provision of a clipper bow, that allows the ship's prow to overhang the underwater protrusion, is clearly safer. The second value of this bow design is simply all about external good looks. It has very much become the trend, principally because it softens the 'block of flats' look of the modern cruise liner, a look that has been further accentuated over recent years by the necessity to install rows and rows of balconies, in keeping with discerning passengers' expectations. For the same reason designers are maintaining a preference for low, streamlined funnels and rounded forward superstructures. Indeed, the twenty-first-century cruise ship flaunts more frontal curves than the average fashion model, photographing better from a forward quarter.

The massive white bows of the *Oasis* and *Allure of the Seas* are copied from Royal Caribbean's previous Finnish buildings, the *Freedom* and *Voyager* classes, the prow, devoid of any crest, extending further ahead than the bulbous bow, abaft of which is an imposing

Two Queens face to face. Viewed over the bows of *QE2*, the 90,049-gross-ton *Queen Victoria*, completed in 2007, contrasts in forward profile if not in tradition.

A Disney character imaginatively emblazoned on the decorative bows of *Disney Wonder*.

array of four thrusters. Most new cruise ships sport little more than a company badge on their bows, although Disney, typically, have shown more imagination with their duo, the *Magic* displaying Sorcerer Mickey whilst the *Wonder* has Steamboat Willie on her forward end. But if curves are the prevailing taste for cruise ship bows, they have disappeared over time from their sides. Tumble home, a natural inward curve of a vessel's sides between waterline and strength deck levels, emanated from wooden built warships of yesteryear on which it was seen as a method of increasing stability. The narrower upper decks would carry the ships' armaments, requiring less space and resultant deck strengthening. A similar principle was later applied to large passenger liners, saving upper deck weight, although compared with their wooden predecessors their tumble home may have been barely detectable. Cunard's two pre-war Queens had tumble home, as did *France* of 1961, but as prefabricated shipbuilding methods replaced the traditionally curvaceous skeleton constructed on a slipway, all new passenger ships, including, *QE2* of 1969, emerged as slab-sided structures.

As cruise liners' bows have become evermore shapely, their sterns, just like the sides of their hulls, have been losing their attractiveness. The styles of ships' rear ends seem to have been changing like ladies' fashions over the years, especially since the days of the intricately designed overhanging counter that featured on early twentieth-century liners such as *Mauretania* and *Titanic*. This style was developed for protecting the stock and rudder above the waterline, but with the introduction of the fully submerged balanced rudder, its use became obsolete, although for aesthetic purposes it continued to appear on ocean passenger vessels for some years. Between the world wars the cruiser stern became the trend, being a far more straightforward design, as was employed on both *Queens*. The stern's above-water slope would be in parallel with the angle of the ship's bow. From the late '50s the refreshingly simple rounded counter stern became popular, appearing on such eminent liners as *Rotterdam* (1959), *Canberra* (1961) and *Queen Elizabeth 2*. This pleasing design was continued with *Royal Princess* (1984), and even Royal Caribbean's *Sovereign*-class trio are blessed with rounded sterns reminiscent of the great *Normandie*, constructed at their Saint Nazaire birthplace.

The uncomplicated rounded counter stern of *QE2*, so easy on the eye.

Whilst several early 'Wedge Ships' retained the rounded look, a new design, the stern of the future, is now with us. The sheer, squared-off transom allows maximum internal space from waterline level upwards, and allows designers the scope to locate public areas that would have been positioned more centrally in earlier ships, at the aft end. Nowadays, thanks to the advent of vibration-free machinery and propulsion, especially the pod system, roll-free stabilisers and the fact that most cruise ships rarely wander from summer-like conditions, passengers find themselves taking their evening meals in expansive two-deck dining rooms that allow starlit views over the ship's stern. Holland America's *Statendam*-class of the '90s were early exponents of this concept, managing to harmonise their sharply cut-off aft ends with pleasing overall external profiles. The *Queen Victoria*'s largest dining room, the Britannia Restaurant, is located aft, as is *Carnival Destiny*'s Universe Dining Room, beneath which are a number of Ocean View cabins with picture windows.

A new trend, as I write, is the apparently seamless continuation of the transom stern up to the top-deck level of the ship's main superstructure. P&O's *Ventura* and *Azura* are examples of this design that maximise interior space to the very extreme, but threaten to extend the 'block of flats' look to the aft end. But nowhere else afloat is there a stern section that compares with that of the two *Oasis*-class ships. With no two-deck dining room in sight, the area is dominated by the unique Aquatheater that takes up three decks and affords superb views of the ship's white-foaming wake from the aft-facing seating that doubles up as a serious sun-worshipping location in the daytime. The transom stern has a gradual slope outwards, down to the waterline, partly matching the inward slope of the bows, rising to Deck 6 from where a further twelve decks tower up to a pair of adjacent streamlined funnels. The shape of the funnels, together with the sweeping curves of Decks 17 and 18, immediately below, that house the twenty-eight two-storey Loft Suites, successfully, in my opinion, offset to a considerable extent the inevitable high-sided appearance of a ship of such magnitude.

As cruise ships become increasingly reminiscent of floating cities, we shall doubtless find it harder to accustom ourselves to their futuristic profiles. Most of us are resistant to change and find it naturally takes time to accept new ideas. A case in point was the external design of the first *Oriana*, launched for Orient Line some fifty years ago. Essentially a development of the company's other post-war liners, *Orcades* (1948), *Oronsay* (1951) and *Orsova* (1954), which all featured a centrally positioned bridge just forward of a high funnel, a design that had become an Orient Line trademark, *Oriana* differed from the smaller trio by having not one but two funnels, both shaped like upturned flower pots. The aft funnel was, in fact, a dummy, acting as housing for the engine room exhausts.

Unfortunately, in addition to these unusual funnels, the appearance of *Oriana*'s high-sided superstructure, designed to accommodate her eleven passenger decks, was met with much criticism from shipping aficionados in her early days. Constructed with much aluminium to avoid top-heaviness, it was repeatedly the target of remarks on 'how top-heavy' she looked! But as we now know, her builders, Vickers Armstrong of Barrow-in-Furness, had created a forerunner to the modern-day cruise liner.

In 1955 British ship-owners Shaw Savill Line placed a liner on round-the-world service whose overall profile would have a profound effect on the shape of passenger ships to come. *Southern Cross* measured only 20,204 gross tons but was a giant of maritime importance by having all her machinery placed aft. Furthermore, she carried no cargo, just like today's cruise vessels, allowing maximum room for her passengers. Four years on, the 37,783-gross-ton *Rotterdam*, that favourite of *QM2* designer Stephen Payne, captured European headlines with her narrow twin uptakes positioned well aft of centre that, like *Oriana*'s unusual design, initially failed to receive universal approval. Nevertheless, the largest Dutch transatlantic liner ever built became a legend of her time, affectionately dubbed the *Grande Dame*, and, following a cruising career under further ownership, has been transformed into her original glory and moored at her homeport of Rotterdam as a static tourist attraction.

The 1960-built *Oriana*, 41,915 gross tons, was criticised for her 'high-sided' appearance.

BUILDING THE BIGGEST

Oriana's P&O running mate, *Canberra*, was a more natural descendant of *Southern Cross* for having her turbo-electric machinery positioned well aft, a curved forward superstructure and a single main mast located immediately abaft her bridge. Great excitement greeted the launching of this new liner that would become as instantly recognisable on the ocean waves as any other passenger ship afloat. She was publicised as 'A ship that changes the future', and nobody could dispute that. I have two vivid recollections of this ship that moved almost seamlessly into cruising whilst *Oriana* struggled to make the transition from liner service.

The legendary *Rotterdam* of 1959, with her rounded counter stern and unique twin uptakes.

It was September 1961 and the annual family tradition in those days was to holiday on England's south coast – normally Bournemouth or the Isle of Wight – incorporating, without fail, a cruise around Southampton Docks. I shall forever be grateful to my parents, both of whom have sadly passed on, for encouraging my natural love of anything connected with ships and the sea from my childhood days, and a version of the following story appears in my previous book *And the Crew Went Too – the £10 Assisted Passage*, and I trust that should you by any chance, dear reader, have read these particular words before, you will stay with me over the next paragraph or so.

In those days the docks would be constantly occupied by passenger liners of various shapes and sizes and, more often than not, one of the two giant Queens would be berthed alongside the Ocean Terminal that had been built in 1950 specifically to serve the famous pair. On that day it was the turn of the *Queen Mary* to take the accolade of being the largest ship in port, and as our small pleasure boat ventured alongside her I felt I could reach out and touch one of the 10,000 rivets driven into her massive hull and superstructure. We moved on, passing two Union-Castle liners preparing for their 4 p.m. departures to South Africa, and then I sensed a hush of anticipation. Our boat seemed to list as all its occupants leaned towards the starboard side to catch their first glimpse of the white-painted vessel that was coming into view; a vessel that somehow seemed to represent all our futures. *Canberra*'s hull bore the scars of a heavy storm off New Zealand. But there she was, safely returned from that long maiden voyage.

Although almost fifty years have passed since that day at Southampton, the experience has remained indelibly in my mind, just as strongly as the sad events of 30 September 1997, the day that *Canberra* arrived home at the end of her final P&O cruise.

Thousands of well-wishers packed Southampton's Mayflower Park, many dressed in T-shirts displaying *Canberra*'s name. It was an eerily misty morning and the wait for the 'Great White Whale' seemed endless until, suddenly, her familiar profile loomed through the haze. Surrounded by a flotilla of small craft, she sailed proudly past us and slowly but surely turned to nestle against the Mayflower Terminal quayside for the very last time. *Canberra* blasted out several notes of acknowledgment on her whistle, and then all was silent. The crowds dispersed, and I stood alone. Was this really the last time I would see her, or had a secret buyer given her a last-minute reprieve? My hopes were dashed a few days later with the news that she had quietly slipped out of Southampton under cover of darkness on her way to a beach at Gadani, Pakistan, to be broken up. The 1992 Amendments to the SOLAS treaty had given ship-owners until 1 October 1997, just one day after *Canberra*'s poignant arrival, to bring their ships in line with new requirements. Much costly work would have had to have been carried out on the ship if her cruising career was to be continued.

Succumbing also to the new rules was *Rotterdam*, but new buyers, under the name of Premier Cruises, were found to keep her in commercial service as the cruise ship *Rembrandt*.

A further update of SOLAS regulations on 1 October 2010 will require that all ships, regardless of their age, are free of almost all combustible materials within their construction.

Among the complex set of rules applying to ocean passenger ships are requirements that there are two means of exit from all atrium levels, low-level lighting systems are installed and fireproof enclosures are fitted around all stairways. The unexpected withdrawal of *Queen Elizabeth 2* from Cunard service in 2008 was undoubtedly due to these regulations, the Carnival group preferring to add a brand-new ship to its subsidiary Cunard fleet in 2010 than put the thirty-nine-year-old cruise liner through yet another multi-million-dollar refurbishment. There is little room for sentiment in the maritime world of today, but at least the celebrated name of *Queen Elizabeth* is being revived with the arrival of a 92,000-gross-ton cruise vessel from Italy.

Taking hull design to a new extreme was the revolutionary twin-hulled SSC (Semi-Submersible Craft) *Radisson Diamond*, seen as a technological breakthrough in passenger ship concept when she was laid down by Finnyards Oy at Rauma in March 1991. Described as a SWATH (Small Waterplane Area Twin Hull) vessel, she would refine a design that has been with us for some time but never attributed to a passenger-carrying ship of her size. By having a deep draught and narrow waterline, she was expected to offer unequalled stability and greatly reduced roll, particularly with her four stabilisers in action.

The disadvantage of the SWATH idea is that it limits the ship's length and cruising speed. Despite having a gross tonnage of 20,295, *Radisson Diamond* was designed to

The brand-new *Canberra* berthed at Southampton in September 1961.

A revolutionary guest at Tilbury. The twin-hulled *Radisson Diamond* created much interest when introduced in 1992.

have a length of only 125m and a top speed of just 14.5 knots. On completion she was introduced to the cruise industry amid a fanfare of expectancy, and I was privileged to be invited to attend her naming ceremony on the River Thames at Greenwich on 28 May 1992. International opera singer Dame Kiri Te Kanawa was asked to perform the christening, and she had the unique task of breaking a bottle of champagne over each of the two hulls.

The 354-passenger *Radisson Diamond* was built for Diamond Cruise Inc., a joint venture that included Radisson Hotels International and Japan's Mitsui OSK Lines, who operated the ship on three- and four-day Caribbean cruises and relatively short Mediterranean itineraries, because of her limitations, and in June 2005 she was acquired by Asia Cruises, owned by Hong Kong casino boss Stanley Ho. The catamaran concept may be fine for small vessels but is yet to prove its worth within the cruising world.

Funnels – a feature of ships that by size, shape, colour and markings permit identification of the vessel's owners or operators, and even the individual vessel itself. For

With a little help from her friends, Dame Kiri Te Kanawa prepares to break a bottle of champagne over one of *Radisson Diamond*'s two hulls.

many years, especially up until the Second World War, funnel designs were predominately uniform; cylindrical structures with straight tops that became increasingly wider, or oval in shape, as time went on. In the earlier days of coal fire engines they were often referred to as smoke stacks (as we have seen in Chapters One and Two) and most had their top portions painted black where soot discolouring would otherwise show, but the change to oil-fired machinery, especially the introduction of diesel engines, and improved technological know-how in ship design and construction has paved the way for more individualistic designs such as on the two *Rotterdams*, *Canberra* and many contemporary cruise ships. There is now an increasing trend to produce ships that are 'eco-friendly', in accordance with the world's long-term attempt to reduce global warming, and their imposing funnels are featured just as much for aesthetic purposes as for distributing the emission flow.

The subject of ships' funnels cannot be allowed to pass without a mention of those wonderful winged structures that are a vital part of the design of Carnival Cruise Line ships. This unique 'look' came about initially by accident but later by deliberate design. In 1968 Canadian Pacific changed its house flag from the familiar red and white checks to a corporate symbol consisting of greens and white. The new logo also appeared on the funnels of all their fleet, which included the *Empress of Canada* that was later laid up at Tilbury awaiting sale. As a money-saving exercise, Carnival adapted the CP logo for its house flag – a red square with a blue circle within a white circle – upon the purchase of *Empress of Canada* in 1972. Renamed *Mardi Gras*, her funnel was repainted with half the house flag's logo, and was also given a red band along the top of her hull. These colours have remained on Carnival Cruise Lines' ships ever since, as has the funnel concept with its sweeping 'gull wing' fins that Joe Farcus designed for their early new build, *Tropicale*.

Some people say that the Carnival design was created in tribute to the SS *France* of 1961, but although the similarity cannot be denied, the story is untrue. But the

Wings across the world. The swept-back Carnival funnel has become instantly recognisable on the cruise circuit.

Distinctive even at twilight, *Norway*'s funnels remained unaltered from her French days.

creation of the *France*'s winged funnels was an inspiration, distinguishing her from her transatlantic competitors of the day and allowing her to retain her individual identity amidst her white-painted Caribbean contemporaries during her lofty days as the cruise ship *Norway*.

Designs of passenger ships' interiors have principally been governed over the years by changes in fashion, fire safety rules and technical know-how. From the Victorian and Edwardian days of ornamental fittings, carved pillars, heavy and dark furnishings and a predomination of wood panelling through to the mid-nineteenth century with its lighter colourings and simpler styles, marine architects have brought to the seas the ever-moving trends of top mainland hotels, adding their own individuality. Now the prevalence of genuine wood is no more, not even the practical Formica of the 1960s, surrendered to the modern-day stringencies of fire safety regulations. Yet, in the twenty-first century, some ship designers are being encouraged to replicate those golden days of liner travel as they endeavour to charm today's cruise passenger with extravagantly chandeliered multi-tiered dining rooms and grand entrance staircases, constructed in non-combustible materials and enhanced by high-tech lighting systems. Among those cruise liners so fashioned are Cunard's *QM2* and the new *Queen Elizabeth* that recalls the early glamour days of her older namesake.

Normandie's semi-circles, applied here to her shopping area frontage.

Lost in time is the once-essential Smoking Room, traditionally of distinguished decor, replaced in some cruise ships by a copy of a gentleman's club (Michael's Club appears universally on all the latest Celebrity Cruises vessels). Other features, pertaining to specific ships, have displayed an element of quirkiness in their design. The unaccountable insistence by the *Normandie*'s designers, for instance, to furnish their ship with everything semi-circular – shops, bars and the Purser's Desk – all reflected the semi-circles of the terraced decks outside. We can only wonder whether the idea would have influenced subsequent liner onboard concepts if *Normandie* had lived a full life.

Whilst the gardens of the *Oasis*-class duo's innovative Central Park 'neighbourhood' can really not be attributed directly to Mewès' wonderful Winter Gardens on the Cunard four-stacker *Aquitania* (he could surely not have foreseen that greenery and flora would one day be displayed aboard an ocean ship on such mammoth scale), the Palm Court/Winter Garden theme has been refined and redeveloped on numerous liners and cruise vessels to this day. It was even a tourist-class amenity on the *Queen Elizabeth* of 1940 and now appears as the Garden Lounge, a beautiful room containing an arched glass

All wickerwork and foliage. One of the pair of Garden Lounges designed for *Aquitania* by Arthur Davis.

The Winter Garden concept was applied even to the tourist class of *Queen Elizabeth* of 1940.

ceiling inspired by the glass houses of the famous Kew Gardens, on the 2010 build of the same name.

The *Oasis* ships' 100m by 19m Central Park is landscaped with flowers, shrubs and trees, the first time real foliage had been extensively planted in 'outdoor' conditions on an ocean-going vessel. Royal Caribbean claim that this area is drawn upon the concept of 'surprise', and passengers will certainly be surprised when some of the tropical trees grow to an expected two-and-a-half decks height. Flowing vines and various ferns trail upwards across four storeys and the park is divided into sections known as The Pergola Garden, featuring Caribbean vegetation, The Sculpture Garden with its artwork and sweet olive trees, The Chess Garden, complete with large-scale pieces and a 'middle pathway' that takes passengers over a bridge spanning a 'river of plants'. Within the expansive Central Park 'neighbourhood' are four separately-themed eating places and a wine bar.

With passenger ship designers becoming ever more creative as they try to stay one step ahead of the changing tastes and expectations of the cruise 'vacationer', it is not at all easy to envisage what the coming years may hold. Could 'gardens at sea' be the next essential cruise liner amenity? Already Celebrity's three 122,000-gross-ton *Solstice*-class ships, the biggest so far built at Meyer Werft's Papenburg Yard, are offering the rare delicacy of a manicured lawn – half an acre of top deck grass. But most importantly, passengers want to feel pampered, just like the voyaging aristocracy of 100 years ago.

They demand facilities that are at least equal to those of luxury hotels ashore. As a result, the ships' spa areas are fast becoming floating resorts in themselves. Most cruise ships have their own thermal suites – steam rooms, saunas and heated relaxation beds – and many have much more. Royal Caribbean's Vitality at Sea Spa and Fitness Centre, as featured on the *Oasis* sisters, has three couple massage suites and seven individual treatment suites. Celebrity Cruises, now a subsidiary of Royal Caribbean International, has been turning up the heat on its competitors for some years and once again lead the way with the their innovative Aquaclass veranda cabins, which provide private access to the ship's spa and their own healthy eating restaurant, on *Celebrity Solstice*, *Equinox* and *Eclipse*, delivered between 2008 and 2010. It seems a far cry from the days of those converted indoor swimming pools on *Queen Elizabeth 2* and *Norway*, once extolled as the best spas facilities afloat. Yet we should hardly be surprised if the passenger ship's life has changed so much over the past thirty years. Indeed, so has the language. Cabins are now staterooms, passengers are known as guests, and cruise lines employ hotel managers and cruise directors.

As cruise liner concepts grow in sophistication and versatility, so do the ships' itineraries. The most 'mega' of vessels, in particular the *Oasis* pair that heavily feature Caribbean themes throughout their onboard amenities, are undoubtedly committed to serving the mass American market, for which they are most economically suited, for many years ahead. Indeed, there are few ports outside that region that can handle such giants, although other parts of the world are making great efforts to stake their claim to a piece of the cruising action. P&O and Princess Cruises have created a steady influx of their vessels into Australian waters over the past five years, including the 'mega' *Princesses* such as *Dawn*, *Sapphire* and *Diamond*, whilst Dubai has produced its own dedicated cruise terminal as it bids for a niche in the sea vacation industry that would endorse its astonishing rise to prominence as a world-class residential and travel location. Costa Cruises, part of the Carnival group, are in the midst of a five-year deal with Dubai's Department of Tourism and Commerce to operate a series of cruises to nearby Middle Eastern ports and, likewise, Royal Caribbean have deployed their 90,090-gross-ton *Brilliance of the Seas* on cruises out of Dubai.

An attractive proposition, in my humble opinion, would be the incorporation of an overnight stay on the converted *QE2* as part of a cruise package based in Dubai. However, as I compile the closing words of this book, *QE2* still lies at nearby Mina Rashid where she finished her Cunard career in November. She technically remains an ocean-going vessel, and former captain Ronald Warwick, who is employed by V-Ships, managers of *QE2* until her refit, is her legal master. Even prosperous Dubai, where the rich and famous own considerable property, is feeling the effect of the worldwide economic depression, although Nakheel, *QE2*'s new owners, insist that conversion work, which includes the controversial removal of her famous funnel in favour of luxury penthouses, will definitely proceed in time. The plans are for her to be moved to the Palm Jumeirah to take up residence at a new QE2 Precinct.

On 24 July 2009 the liner entered Dubai Dry Dock for external refurbishment when it was announced that she would be sailing for Cape Town, where Nakheel have marine development interests, to act as a floating hotel for eighteen months, encompassing the 2010football World Cup Finals. It is envisaged that this will provide welcome funds for her eventual conversion at Dubai. I know that I express my hope on behalf of ship-lovers across the world that the great liner finds a permanent home before too long and is not consigned to the scrap yard.

As for the cruise industry as a whole, the future remains optimistic, even exciting, as Norwegian Cruise Line take delivery of a 150,000-gross-ton megaship, *Norwegian Epic*, in 2010, the largest ship to be built at Saint Nazaire, Carnival Cruise Lines accept *Carnival Dream* which, at 130,000 gross tons, is their biggest cruise liner so far, from Italian builders Fincantieri, and a second resident ship, this time exceeding 100,000 gross tons, is at planning stage. But as the economic downturn continues to hang over the horizon, cruise operators need to tread carefully to avoid finding themselves in deep water. In 2009 Royal Caribbean declined options for a fourth *Freedom*-class ship and a third *Oasis* vessel. So, for the foreseeable future at least, those two giant floating cities, *Oasis of the Seas* and *Allure of the Seas*, will hold the accolade of being the biggest passenger ships of all time.

INDEX

INDEX